Design of Sustainable Product Life Cycles

Jörg Niemann · Serge Tichkiewitch ·
Engelbert Westkämper (Eds.)

Design of Sustainable
Product Life Cycles

 Springer

Dr.-Ing. Dipl.-Wirt. Ing. Jörg Niemann
Institut für Industrielle Fertigung
und Fabrikbetrieb
Fraunhofer Institut für Produktionstechnik
und Automatisierung (IPA)
Nobelstraße 12
70569 Stuttgart
Germany
jon@iff.uni-stuttgart.de

Prof. Dr. Serge Tichkiewitch
G-SCOP INPG
Domaine Universitaire,
BP 53
38041 Grenoble Cedex 9
France
serge.tichkiewitch@inpg.fr

Prof. Dr.-Ing. Engelbert Westkämper
Institut für Industrielle Fertigung
und Fabrikbetrieb
Fraunhofer Institut
für Produktionstechnik und
Automatisierung (IPA)
Nobelstraße 12
70569 Stuttgart
Germany
wke@iff.uni-stuttgart.de

ISBN: 978-3-540-79081-5   e-ISBN: 978-3-540-79083-9

Library of Congress Control Number: 2008936041

Cover design: eStudio Calamar S.L.

Printed on acid-free paper

9 8 7 6 5 4 3 2 1

springer.com

Preface

"Improved performance while lowering environmental impact"

Thomas Bittner
is Head of Service and Fullservice
at ABB Automation GmbH, Germany

Up till recently, improved performance in the process industry has always been characterised by the optimisation of the process control implemented. This involved higher costs to continuously adapt systems in order to keep pace with market change drivers. However, volatile customer requirements call for machines with increasingly shorter cycle times and smaller lot sizes. As a result, (short-term orientated) process optimisation alone is no longer a guarantee for success in today's world. In order to achieve sustained competitive advantages in the future, companies will have to be capable of exploiting in the long-term the as-yet untapped productivity potentials of the assets they implement over their entire life cycle.

In the context of future-proof life cycle management, there is a need to accompany the customer's production system with intelligent evolution strategies throughout its life cycle and to optimise it in alignment with technical/technological advancements according to the situation and requirements prevailing. Thus, the management of the different life cycles of components within a system is especially important. Here, the design of sustainable product life cycles means especially the pro-active, long-term protection and safeguarding of the knowledge of the process sequence control which has been implemented by the customer in the system over the years. This process can only become successful through continuous system management over the entire life cycle of the customer's plant/machine.

Consequently, we are developing resource- and environmentally-friendly solutions to enable our customers to use electrical power efficiently, increase their industrial productivity and lower environmental impact in a lasting way. This not only allows plant manufacturers to use their investments in a more sustained and efficient way, it also creates the chance to form and consolidate enduring customer relationships with the aid of tailormade, life cycle-orientated service strategies which will be reflected in long-term and commercially successful partnerships.

Thomas Bittner

Preface

"Systematic approach to life cycle management"

Dr. Frank Bünting
is life cycle costing expert
at the VDMA

Every company talks about forming long-term, intensive relationships with its customers. The link connecting the customer to the manufacturer is the product which the customer doesn't only buy but also uses for a planned length of time. As a result, the behaviour of the product during this period of time plays a significant role in the success of a long-term relationship.

Therefore, manufacturers are well-advised to systematically take the life cycle of their products into consideration. This because characteristics such as mean time between failure (MTBF), meantime to repair (MTTR), availability and product lifetime are no longer empty words but rather figures which are being included more and more frequently in contract negotiations. After all, for many customers, it isn't just the purchase price which counts any more. Today, a whole range of cost considerations are made with the result that parameters such as energy consumption, material consumption and the cost of replacement parts are becoming increasingly important when making investment decisions. For example, Daimler only places orders for investment goods if an agreement about life cycle costs has been made.

One of the problems associated with predicting life cycle costs is the comparison of information. An important contribution towards standardising LCC approaches and thus making them suitable for practical use has been made by the VDMA Specification 34160: "Forecasting Model for Lifecycle Costs of Machines and Plants". The specification defines a calculation model which can be individually adapted to the needs of the customer and the supplier without losing the ability to compare. Many companies now implement this model for forecasting purposes or as a tendering requirement.

Dr. Frank Bünting

Preface

"Thinking in product life cycles"

Prof. Dr.-Ing. Prof. E.h. Dr.-Ing. E.h. Dr. h.c. mult.
Engelbert Westkämper
is managing director of the Fraunhofer Institute for Manufacturing
Engineering and Automation (IPA) and director of the Institute of Industrial
Manufacturing and Management (IFF)University of Stuttgart, Germany

Product life cycle management is a subject which is increasingly gaining importance in management circles and considers the complete life cycle of products from the phases of development, manufacture, sales and service right up to recycling. In the process, the sustainable design of product life cycles guarantees targeted customer orientation and enduring customer loyalty throughout all the stages of a product's life.

A significant demand which will be placed on mechanical engineering in the future will be to extract the maximum benefit from all industrially manufactured products over their entire life cycle. Taking the fact of limited natural resources such as energy and materials into account which will pose considerable problems for the coming generation, the aim must be to achieve maximum product efficiency using a minimum of consumed, non-renewable resources.

With the latest approaches, goods remain in the manufacturer's network for their total lifetime with the result that the manufacturer also becomes responsible for the life cycle management of his products. Here, a central function is the continuous preparation and processing of life-cycle related product data "from the cradle to the grave" to ensure permanent exploitation of the resources utilised and maximisation of benefits.

Following this new paradigm, we have started to place the life cycle of technical products at the core of our developments.

In order to achieve this, we are relying on the central competencies of our engineers to develop technically and economically viable solutions as well as accompanying life cycle information technology possessing new perspectives in this regard.

E. Westkämper

Foreword

Companies in businesses orientated towards the long-term are no longer able to limit themselves just to the economic success of their commercial operations. Today they also have to consider the ecological and social consequences of their business activities.

Subprocesses associated with the development, manufacture, usage and disposal of products only result in suboptimum value-adding processes. However, the aim should be to achieve an overall optimum.

The concept of life cycle management deals with this problem and takes product life cycles into account in a comprehensive philosophy. The core of this concept is the implementation of an integrated planning approach when considering product life cycles, creating long-term horizons to gain maximum usage of products over their entire life cycles. In order to operate and manage technical products and equipment over such a long period of time, in the future all those involved in the development, manufacture, usage and disposal of a product - i.e. all life cycle partners from the "cradle to the grave" - will have to work together.

These aspects contain new potentials regarding added value and are exceptionally important to the manufacturers and operators of high-quality investment goods. The maximum exploitation of a product represents product profit and is therefore a new paradigm shift because it breaks away from traditional paradigms regarding growth and resource optimisation. With this innovative approach, it is the benefit of a product for the customer, or respectively the long-term business relationship and thus the complete life cycle of a product which is in the foreground and the optimisation of the separate phases along the way fades into the background.

If the thought spanning the total management of product life cycles is further pursued in a consistent way, in the near future the manufacturers of production equipment may only sell the usage of the systems they produce. With the aid of a network of other partners and specialists, they will look after the systems being utilized by their customers and retain responsibility for the systems they supply. This approach is already being successfully implemented today for painting plants in the automotive industry or for printers, for example. As a result, future equipment manufacturers will become suppliers and producers at the same time.

This book concerns itself precisely with this change in paradigm and starts by describing relevant sub-aspects in which the philosophy of "thinking in terms of product life cycles" plays a major role. In the process, the book presents common tools and innovative methods for modelling and optimising life cycles. Subsequent chapters discuss approaches and methods for designing products with a view to their life cycles and for life cycle information support in order to continuously optimise life cycles. In further chapters, ways and effects of product-supporting services are presented as well as concepts for life cycle orientated cost and analysis instruments. Particular attention is paid to the consideration of networks which link customers and suppliers throughout the life cycle of a product. These networks not only represent the basis for commercial success but are also door-openers for additional services. In this way, products mutate into vehicles with which additional turnover potentials can be exploited.

In another chapter, the approaches discussed are combined to form a new integrated method orientated towards the long-term which should help the reader to develop and successfully manage sustainable life cycles for products based on the new paradigm.

Life cycle design is understood as being "the development" (planning, calculation, definition, drawing) of a holistic concept for the entire life cycle of a product. Life cycle design means once-only planning during the conception phase of a product in which the pathway of a product is determined throughout its entire life cycle. It includes, for example, the planning of possible services for a product during its utilization phase, material recycling methods, the possible reuse of certain parts, the organisation of recycling logistics and the possible later use of a product. Thus it is a conceptual pre-design of all later activities over the total product life cycle.

The book addresses professionals as well as researchers and students from the field of life cycle management. Heading for life cycle excellence, practitioners and researchers alike will benefit from the coverage of comprehensive methods and the presentation of various examples from industry.

This monograph is the result of four years of work by the VRL-KCiP, the European Knowledge Community for Holistic Production. It is a joint venture undertaken by twenty-four internationally well-renowned research institutions from fifteen different countries developed from a Network of Excellence in the 6th Framework Programme of the European Commission. In accordance with the KCiP charter to gain advantages through stronger networking, all contributing partners were involved in writing the book in parallel. As opposed to most book-writing approaches where single authors are involved in writing it chapter by chapter, this book is the common work and common understanding of ALL authors who elaborated the ENTIRE contents. As a result, the elaboration process became highly dynamic through the permanently-triggering discussions and improvements made to the manuscript.

On completion of the European Commission programme in 2008, the VRL-KCiP will continue to operate under the name EMIRAcle. To foster holistic production research, EMIRAcle accesses the know-how of its partners who excel in the various aspects of production. Thus it will become possible to combine an unprecedented range of competencies and to transfer the knowledge generated to industry. The cooperation between research and industry raises the researcher's awareness of "real-life" problems and tasks encountered in industry as well as providing industry with the latest joint research results.

The editors would like to thank the large number of authors who contributed to this innovative European-wide approach. This book was only possible by using all intellectual sources from the network. The experiences gained while compiling this monograph have paved the new way for thinking in networks and collaborative partnerships.

The editors would especially like to take this opportunity to thank Thomas Bittner, ABB Service Automation and Dr. Frank Bünting from VDMA for their prefaces written from an industrial perspective. Special thanks go to Max Dinkelmann, who - with his distinguished expertise and ideas concerning the development of a standardised method for life cycle design - made a major contribution towards the completion of the book. The editors would also like to thank Helen Schliesser for her final proof reading and for giving valuable input regarding the unification of different "European terminologies" in the manuscript review. Finally, thanks also go to Dr.-Ing. Boris Gebhard from Springer Verlag for his cordial support while the book was in the process of being written.

September 2008, for the editors,

Jörg Niemann

List of authors

ALDINGER, Lars, Dipl.-Ing.
Institute of Industrial Manufacturing and Management (IFF), University of
Stuttgart, Allmandring 35, 70569 Stuttgart, Germany

ALZAGA, Aitor
Fundacion Tekniker, Avda. Otaola, 20, 20600 Eibar, Spain

BAGULEY, Paul
University of Durham, Net Park Research Institute, Joseph Swan Road,
Sedgefield, TS21 3FB Durham, United Kingdom

BITTNER, Thomas
Dipl.-Kfm., ABB Automation GmbH, Kallstadter Straße 1, 68309 Mannheim,
Germany

BOËR, Claudio
Prof., Istituto Tecnologie Industriali e Automazione, Via Bassini, 15,
20133 Milan, Italy

BOSSIN Donna
M.A., Technion, Israel Institute of Technology, Department of Mechanical
Engineering, 32000 Haifa

BRAMLEY, Alan
Prof. Dr., University of Bath, Claverton Down, BA2 7AY, Bath, United Kingdom

BRISSAUD, Daniel
Prof. Dr., INPG, Institut National Polytechnique de Grenoble, Domaine
Universitaire BP 53, 38041 Grenoble, Cedex 9, France

BÜNTING, Frank
Dr. rer. pol., Verband Deutscher Maschinen- und Anlagenbau e.V.), VDMA,
Lyoner Straße 18, 60528 Frankfurt am Main

BUFARDI, Ahmed
Dr., EPFL Swiss Federal Institute of Technology, Lausanne, Chemin des
Machines, CH-1015 Ecublens, Switzerland

CHRYSSOLOURIS, George
Prof. Dr.-Ing., Laboratory for Manufacturing Systems and Automation (LMS),
Department of Mechanical Engineering and Aeronautics, University of Patras,
Patras 26110, Greece

COLLEDANI, Marcello
Politecnico di Milano, Dipartimento di Meccanica, Via Bonardi 9, 20133 Milan, Italy

DINKELMANN, Max
Institute of Industrial Manufacturing and Management (IFF), University of Stuttgart, Allmandring 35, 70569 Stuttgart, Germany

DORI, Dov
Prof., Technion Israel Institute of Technology, Department of Industrial Engineering &Management, 32000 Haifa, Israel

DRAGHICI, Gheorge
Prof. Dr., "Politehnica" University of Timisoara, P-ta Victoriei Nr. 2, 300222 Timisoara, Romania

DRAGHICI, Anca
Dr. Ing., "Politehnica" University of Timisoara, P-ta Victoriei Nr. 2, 300222 Timisoara, Romania

DU PREEZ, Nicolaas Deetlefs
Prof., University of Stellenbosch, Banhoek road, 7602 Stellenbosch, South Afrika

ENPARANTZA, Rafael
Dr., Fundacion Tekniker, Avda. Otaola, 20, 20600 Eibar, Spain

FISCHER, Anath
Prof., Technion Israel Institute of Technology Department of Mechanical Engineering, 32000 Haifa, Israel

GIESS, Matt
Dr., University of Bath, Claverton Down, BA2 7AY, Bath, UK

GROZAV, Ion
Dr. Ing., "Politehnica" University of Timisoara, P-ta Victoriei Nr.2, 300222 Timisoara, Romania

HAAG, Holger
Dipl.-Ing., Institut für Industrielle Fertigung und Fabrikbetrieb, Allmandring 35, 70569 Stuttgart, Germany

HAYKA, Haygazun
Dr.-Ing., Fraunhofer IPK, Pascalstraße 8-9, 10587 Berlin, Germany

ILIE-ZUDOR, Elizabeth
Dr., Hungarian Academy of Sciences Computer and Automation Research Institute, Kende u. 13-17, 1111 Budapest, Hungary

JOVANE, Francesco
Prof., ITIA-CNR Istituto Tecnologie Industriali e Automazione, Viale Lombardia, 20/A, 20131 Milan, Italy

KALS, Hubert
Prof. Dr., University of Twente, Post Box 217, 7500 AE Enschede, The Netherlands

KIND, Christian
Dipl.-Ing., FhG/IPK, Fraunhofer Institut für Produktionsanlagen und Konstruktionstechnik, Pascalstr. 8-9, 10587 Berlin, Germany

KJELLBERG, Torsten
Prof. Dr., KTH, Royal Institute of Technology, Brinellvägen 66, SE-10044 Stockholm, Sweden

KOMOTO, Hitoshi
Dipl.-Ing. (Mach), 1. Design for Sustainability, Department of Design Engineering, Faculty of Industrial Design Engineering, Landbergstraat 15, 2628 CE Delft, The Netherlands.

KRAUSE, Frank-Lothar
Prof. Dr.-Ing., FhG/IPK, Fraunhofer Institut für Produktionsanlagen und Konstruktionstechnik, Pascalstr. 8-9, 10587 Berlin, Germany

LUTTERS, Eric
Dr.Ir, University of Twente, Post Box 217, 7500 AE Enschede, The Netherlands

MAROPOULOS, Paul George
Prof., University of Durham, Old Shire Hall, DH1 3HP, Durham, United Kingdom

MATTHEWS, Peter
Dr., University of Durham, South Road, DH1 3LE, Durham, United Kingdom

MAVRIKIOS, Dimitris
Dr., Laboratory for Manufacturing Systems and Automation (LMS), Department of Mechanical Engineering and Aeronautics, University of Patras, Patras 26110, Greece

MOLCHO, Gila
Dr., Technion Israel Institute of Technology, Department of Mechanical
Engineering, 32000 Haifa, Israel

MONOSTORI, László
Prof. Dr., Hungarian Academy of Sciences, Computer and Automation Research
Institute, Kende u. 13-17, 1111 Budapest, Hungary

NIEMANN, Jörg
Dr.-Ing. Dipl.-Wirt. Ing., Honorary Professor TU Cluj-Napoca (Rumania),
ABB Automation GmbH, Oberhausener Sr. 33, 40472 Ratingen, Germany

NOEL, Frédéric
Prof. Dr., INPG-UJF, Université Joseph Fourier, Domaine Universitaire, BP 53,
38041 Grenoble, Cedex 9, France

NYQVIST, Olof
KTH Royal Institute of Technology, Brinellvägen 66, SE-10044, Stockholm,
Sweden

PARIS, Henri
Dr., INPG-UJF, Université Joseph Fourier, Domaine Universitaire, BP 53, 38041
Grenoble, Cedex 9, France

ROGSTRAND, Victoria
KTH Royal Institute of Technology, Brinellvägen 66, SE-10044 Stockholm,
Sweden

ROMERO, Ricardo
Fundacion Tekniker, Avda, Otaola, 20, 20600 Eibar, Spain

ROTHENBURG, Uwe
Dipl.-Ing., FhG/IPK, Fraunhofer Institut für Produktionsanlagen und
Konstruktionstechnik, Pascalstr. 8-9, 10587 Berlin, Germany

ROUCOULES, Lionel
Dr., University of Technology of Troyes, 12 rue Marie Curie, 10010 Troyes,
France

SACCO, Marco, Dr.
Istituto Tecnologie Industriali e Automazione, Via Bassini, 15, 20133 Milan, Italy

SALONITIS, Konstantinos, Dr.,
Laboratory for Manufacturing Systems and Automation (LMS), Department of Mechanical Engineering and Aeronautics, University of Patras, Patras 26110, Greece

SCHNEOR, Ronit
M.Sc., Technion, Israel Institute of Technology, Department of Mechanical Engineering, 32000 Haifa, Israel

SHPITALNI, Moshe
Prof. Dr., Technion Israel Institute of Technology, Department of Mechanical Engineering, 32000 Haifa, Israel

SHTUB, Avraham
Prof., Technion Israel Institute of Technology, Department of Industrial Engineering &Management, 32000 Haifa, Israel

SIVARD, Gunilla
Dr. Ing., KTH Royal Institute of Technology, Brinellvägen 66, SE-10044 Stockholm, Sweden

STAVROPOULOS, Panagiotis
Dr., Laboratory for Manufacturing Systems and Automation (LMS), Department of Mechanical Engineering and Aeronautics, University of Patras, Patras 26110, Greece

STOLZ, Marcus
Dr.-Ing., Klingelnberg AG, Turbinenstrasse 17, 8023 Zürich, Switzerland

TE RIELE, Freek L.S.
University of Twente, Department of Mechanical Engineering, Biomechanical Engineering Group, BMTI, P.O. Box 217, 7500 AE Enschede, The Netherlands

TICHKIEWITCH, Serge
Prof. Dr., G-SCOP INPG, Domaine Universitaire, BP 53, 38041 Grenoble, Cedex 9, France

TOLIO, Tullio
Prof., Politecnico di Milano, Dipartimento di Meccanica Via Bonardi, 9, 20133 Milan, Italy

TOMIYAMA, Tetsuo
Prof. Dr., Intelligent Mechanical Systems, Department of Biomechanical Engineering, Faculty of Mechanical, Maritime and Materials Engineering, Delft University of Technology, Mekelweg 2, 2628 CD Delft, The Netherlands

TOXOPEUS, Marten
Ir., University of Twente, Post Box 217, 7500 AE Enschede, The Netherlands

TURC, Cristian
Dr. Ing., "Politehnica" University of Timisoara, P-ta Victoriei Nr.2, 300222 Timisoara, Romania

URGO, Marcello
Politecnico di Milano, Dipartimento di Meccanica Via Bonardi, 9, 20133 Milan, Italy

VAN DRIEL, Otto P.
Ir., Technische Universiteit Eindhoven, Den Dolech 2, Pav C.03, 5612 AZ Eindhoven, The Netherlands

VAN HOUTEN, Fred J.A.M.
Prof., University of Twente, Post Box 217, 7500 AE Enschede, The Netherlands

VÁNCZA, József, Dr.
Hungarian Academy of Sciences, Computer and Automation Research Institute, Kende u. 13-17, 1111 Budapest, Hungary

WESTKÄMPER, Engelbert
Prof. Dr.-Ing. Prof. E.h. Dr.-Ing. E.h. Dr. h.c. mult., Institute of Industrial Manufacturing and Management (IFF), University of Stuttgart, Fraunhofer IPA, Stuttgart, Allmandring 35, 70569 Stuttgart, Germany

XIROUCHAKIS, Paul
Prof. Dr., EPFL Swiss Federal Institute of Technology, Lausanne, Switzerland

Contents

List of abbreviations

ARIS	Architecture of Integrated Information Systems
BLOT	Build-Lease-Operate-Transfer
BOL	Beginning Of Life
BOO	Build-Operate-Own
BOT	Built-Operate-Transfer
BPML	Business Process Modelling Language
CALS	Continuous Acquisition and Life Cycle Support
CIM	Computer-Integrated Manufacturing
DIN	Deutsches Institut für Normung e.V.
DLA	Defence Logistics Agency
DLT	Digital Life cycle Technology
DRed	Design Rationale Editor
EA	Environmental Assessment
Ed.	Editors
EN	European Norm
EOL	End Of Life
EPC	Event Process Chain
et al.	et alii
EU	European Union
ISO	International Organisation for Standardisation
JECPO	Joint Electronic Commerce Program Office
KM	Knowledge Management
LCA	Life Cycle Assessment
LCC	Life Cycle Costing
LCC	Life cycle Cost/Costing
LCS	Life cycle Systems
LCU	Life Cycle Unit
MOL	Middle Of Life
MTBF	Mean Time Between Failures
NPV	Net Present Value
OEE	Overall Equipment Effectiveness
OEM	Original Equipment Manufacturer

OPD	Object-Process Diagram
OPL	Object-Process Language
OPM	Object Process Methodology
OPM	Object Process Methodology
PCM	Product Condition Model
PDML	Product Data Markup Language
PEIDs	Product-Embedded Information Devices
PLC	Programmable Logic Controls
PLCS	Product Life Cycle Support
PLM	Product Life Cycle Management
PM	Product Model
PMPP	Post Mass Production Paradigm
PPC	Production Planning and Control
PSCD	Product and Service Co-Design
PSL	Process Specification Language
PSL	Process Specification Language
PSS	Product-Service Systems
RAM	Reliability, Availability and Maintainability
RFID	Radio Frequency Identification tag
STEP	STandard for theExchange of Product model data
UML	Unified Modelling Language
VDI	Verein Deutscher Ingenieure
VDMA	Verband Deutscher Maschinen- und Anlagenbau e.V.
WEEE	Waste Electrical and Electronic Equipment
XML	Extensible Markup Language
XPDL	Process Definition Language

1 Introduction

1.1 The new paradigm

Industrial manufacturing and the consumption of technical products have led to a dramatic depletion of natural resources and an increasing strain on the environment due to emissions. Society's heightened ecological awareness is taking effect, with the result that more and more companies are publicly committing themselves to environmental protection. In the process, laws and requirements are bringing about a change in the management of resources. Many companies now recognise the fact that they can make cost savings by encapsulating critical technical processes and handling problem materials more frugally. Today, this development is leading to a rediscovery of the product life cycle. In consequence, this development is also strengthening the sustainability sought by politics and society with regard to responsible commercial trading. Commercial sustainability means that all trade should be orientated towards maintaining all resources. The question at the core of manufacturing is how to achieve overall value creation with one product over its entire lifetime by taking life cycle management into account. Consequently, a change in strategies has taken place which not only takes economical aims but also ecological and societal aspects into consideration in the design and utilisation of technical products. (Figure 1.1)

Manufacturers have to accept more and more responsibility for the usability of their technical products and for the consequences of usage. However, many companies only follow statutory general conditions in pre-sales and after-sales in order to avoid losing their markets. There is a general impression that the cost-benefit ratio, especially in after-sales business, is insufficient. This also applies to industrial recycling. One main factor is the availability of actual information about the products and a lack of synergy between final assembly and after-sales operations. (Westkämper et al. 2000), (Anderl et al. 1997), (Alting et al. 1997)

The development of modern products is being decisively influenced by the application of technologies contributing towards increased efficiency. Products are becoming complex highly-integrated systems with internal technical intelligence enabling the user to implement them reliably, economically and successfully even in the fringe ranges of technology. As a result, business strategies are aiming more and more at perfecting technical systems, optimising product usage and maximising added value over the entire lifetime of a product. In this context, the total management of product life cycles coupled with the integration of information and communications systems is becoming a key success factor for industrial companies. (Seliger 1997), (Zülch et al. 1997)

When manufacturing technical products, industrial companies generally direct their strategies towards economic targets. Their main business lies in developing, producing and operating products either for individual customers or for complete sectors of the market. Service and maintenance are considered by many companies to be necessary in order to achieve lasting business relationships with customers.

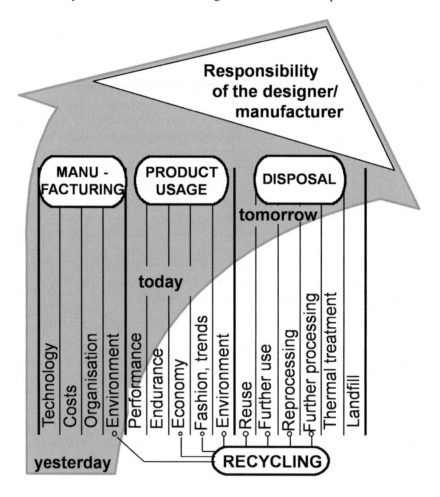

Fig. 1.1: Increased responsibility over the entire product life cycle.

Several studies indicate that the role of these services will change from being a product-accompanying service to becoming the main *revenue driver*. This means that the original product itself will turn into a vehicle (platform) to sell such services as *main business*. (Rifkin 2000), (Hoeck 2005)

Consequently, industrial manufacturing companies are increasingly concentrating their businesses on engineering, assembly and services. They are following new paradigms in order to add value through customer orientation, system management and services during the lifetime of their products. The machine manufacturing industry and other industrial fields such as the automobile industry have reduced their own capacities down to the main or core technologies and final assembly. Parts and components are manufactured by suppliers or specialised companies. Profit is increasingly becoming a result of business operations in design, engineering, final assembly and service. These phases of production are the core competencies of companies which produce strong market or customer-orientated products and add value during a product's life cycle. (Feldman 2002), (Westkämper and Niemann 2006)

The functionalities of products are defined in the processes of design and engineering. The functionality of products and their specific or characteristic properties for usage are determined (as built) or altered by assembling, maintaining and disassembling real configurations,. In the usage phase, special know-how regarding design and characteristic properties is required, such as specific process knowledge for optimising utilisation and performance. Increasing technical complexity is promoting product-near services and manufacturer assistance. This brings about new business models for marketing only the functionality of capital-intensive products rather than selling the products themselves.

There is a new paradigm behind these tendencies: in order to add value and maximise utilisation, products are linked in the manufacturer's network from the beginning right up to the end (see Figure 1.2). In order to realise this paradigm, manufacturers need life cycle management (LCM) systems, tools and technologies. The concentration of all processes into the total life cycle of a product and the optimisation of usage of each single technical product can be described as a new paradigm. Seen from a global point or macro-economical point of view, this is only logical. Seen from an operational or micro-economical point of view, it is proving difficult to initiate such strategies. This is because fundamental structural changes are required in products as well as in organisations and production technologies and also that the economic benefits involved are either uncertain or associated with risks. (Westkämper et al. 2000), (Alting et al. 1997)

Additionally, locally optimised product life cycles (i.e. optimisation of individual processes may not exhibit superior performance globally from multi-objective perspectives. Therefore, the performance of product life cycles needs to be evaluated from holistic and multi-objective perspectives.

However, there is a futuristic vision in the life cycle management of optimising the total exploitation of each product and reducing environmental impact to a minimum. In reality, the different types of products need to be taken into account individually. For some products, it makes economic sense to link them to the manufacturer's network. If the futuristic vision is followed that all machines and high-quality technical products remain in the manufacturer's information network, the Internet will attain a central importance in total life cycle management. (Niemann 2003), (Niemann and Westkämper 2004)

The strategies followed by companies are significantly dependent upon the type of product involved. In a preliminary classification, three categories with varying time scales and strategies may be defined. The first category is goods with a short lifetime and a low product value or complexity. Such non-durable technical consumer goods are usually mass-produced and manufactured in large series. Here the main emphasis of life cycle management is placed on the rational organisation of services, marketing and product recycling techniques. Robust techniques can be used for recycling due to the fact that the added value profit is low in relation to the value of the product. The second category is assigned to series products with a limited number of variants. Life cycle management for these products includes services and maintenance as well as industrial recycling and the partial reuse of parts and components. The third category is reserved for high-quality capital goods. The main emphases here are on maximum utilisation strategies, maintaining performance and additional added value in the field of after-sales. Industrial recycling only plays a minor economic role in this category of products. (Brussel and Valckenaers 1999)

Technical products are linked in the manufacturers' network (Internet) to....

Internet – standards
- Network
- Addressing (URL)
- Services
- XML

- Co-operate in engineering and manufacturing on global standards

- Support the customer in all requests

- Optimise product utilisation (tuning) using best practice methods

- Add value through the use of teleservice, teleoperations, reconfiguration, reuse, remanufacturing, recycling

- Manage the total life cycle of specific products

Fig. 1.2: The vision of life cycle management.

A forward-looking life cycle plan for the product is one example of a maximum utilisation strategy. On completion of the usage phase, the owner faces the alternatives of either scrapping/ recycling the product or of upgrading it. Through upgrading, the product is transformed so that it obtains a new operational status reflected in new product functions. Specific software or hardware modifications are carried out on the product to equip it with advanced, extended or new functional features in comparison with its original condition. Consequently, the product can be improved, extended or utilised to perform completely new tasks. Through upgrading, a product almost starts a new life (Figure 1.3).

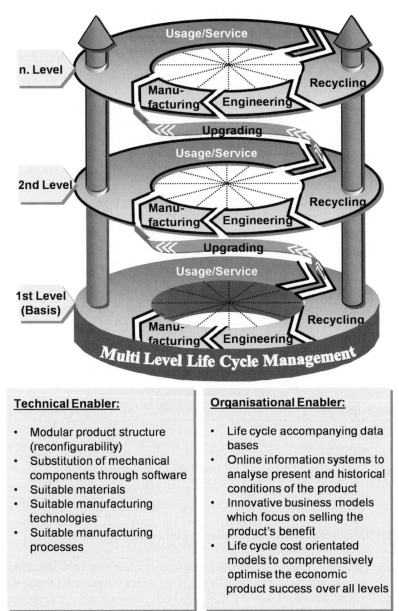

Technical Enabler:

- Modular product structure (reconfigurability)
- Substitution of mechanical components through software
- Suitable materials
- Suitable manufacturing technologies
- Suitable manufacturing processes

Organisational Enabler:

- Life cycle accompanying data bases
- Online information systems to analyse present and historical conditions of the product
- Innovative business models which focus on selling the product's benefit
- Life cycle cost orientated models to comprehensively optimise the economic product success over all levels

© IFF, Westkämper, Niemann

Fig. 1.3: Products have several lives. (Niemann 2003)

However, upgrading is not always possible due to either technical or economic circumstances. In order to be able to upgrade at a later point in time, far-sighted product planning is required which commences in the product engineering stage. In this early phase of development, the fundamental product features - including later modification possibilities - are fixed. Numerous technical and organisational measures decide whether a product can be successfully transformed to attain another level.

From a technical point of view, the modular design of a product's construction is of particular importance. Modular product design in accordance with the laws of system technology enables the variable and economically-viable re-design of a product throughout its entire lifetime. If the fact is taken into consideration that a product may be modified many times over or even altered completely during its lifetime, such product constructions not only bring about advantages for product maintenance but also create enormous potentials. The increasing substitution of mechanical components with software also supports the short-term usage of a product for variable task assignments. Retrofitting times can be shortened due to the fact that modified software can be installed much faster than hardware components can be exchanged. (Westkämper 2000)

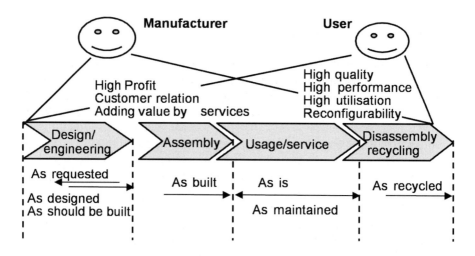

Fig. 1.4: Views of manufacturers and users on the life cycle of technical products.

From a technical point of view, product optimisation can be supported using lifelong data acquisition. Data-logging enables the behaviour of a product to be statistically analysed or products and processes to be monitoring online. The data obtained using this method are evaluated according to specific criteria and discloses optimisation potentials. This permits machines to be completely controlled with the result that, in the future, not only will it be possible to perform technical optimisation but also to take economical factors into consideration and to carry out far-sighted planning thanks to the availability of "real" machine data. Life cycle simulation techniques also enable us to predict product behaviour even in the early phases of the design process. Such real machine data dynamically improve the life cycle model used in life cycle simulation. Up till now, conventional manufacturing paradigms have focused on profit aspects associated with manufacturing and selling products to the end-customer. The new paradigm takes into account the life cycle of technical products and the optimisation of value and benefits during the phases of engineering, assembly, service, maintenance and disassembly. The objective is to reduce environmental losses and to fulfil public or governmental restrictions over the life cycle. (Zülch et al. 1997), (Alting and Jorgensen 1993), (Fullana and Puig 1997), (Kondoh et al. 1998), (Sheng et al. 1998), (Steinhilper 1998)

Following the new paradigm of optimisation and added value over the total life of products, a structural change in the relationship between the manufacturer and the user will take place. Both have different views regarding the same business processes in the life of products, as shown in Figure 1.4.

Different views held about the same product are the result of industrial developments in the 21st century. In the future, the holistic view will offer new ages of manufacturing.

1.1.1 Manufacturer's viewpoint

In general, the life cycle of products can be divided into the phases of design and engineering, manufacturing, assembly, usage, service, disassembly and recycling. Further dimensions are defined in (Seliger 1997), (Westkämper et al. 2000), (Alting and Jorgensen 1993), (Fullana and Puig 1997), (Alting and Legart 1995), (Harjula et al. 1996), (Blanchard 1978), (Niemann 2003) and depend on the specific structure of products and production. The main objective is to fulfil markets and customer requirements to ensure the efficient utilisation of manufacturing resources. The new view adds value in the usage and recycling phases as a result of customer-related services including maintenance and disassembly for reconfiguration, reuse and recycling. More than ever before, this view of the usage and recycling phases makes it indispensable to take into account the various aspects of life cycle design and engineering or the capability of systems to be assembled, disassembled and diagnosed in all phases, especially in that of usage.

It is also necessary to describe the architecture of a product which, in effect, is a mixture of goods and services. Using a model of the integrated architecture, interdependencies between goods and services can be managed more easily because it clarifies how various parts contribute to realising a function (Sivard 2003). As mentioned before, the early phases of the manufacturing process are mostly outsourced to suppliers. Therefore, it is necessary to consider the relationship between manufacturers and suppliers from an economical and environmental perspective. This creates profitable product-orientated services throughout all operations by supporting the diagnostics of actual features, as well as the partial disassembly and assembly for reconfiguration or upgrading and the final disassembly for recycling.

1.1.2 Customer's viewpoint

Customers are generally interested in achieving high product utilisation in the usage phase at the lowest cost, even if this demands that processes be changed. This requires flexible manufacturing systems which provide guaranteed process performance and require minimal set-up times and costs. The high efficiency of the usage of complex technical products depends on specific skills and know-how concerned with the details of machines, mechatronic components, software and process optimisation. These costs can be overcome by using specific skilled services and assistance or support provided by manufacturers. Users prefer buying specialised services to reduce the fixed costs of products as well as the costs of inspection, maintenance and reconfiguration or upgrading.

The economic efficiency of capital-intensive products in industrial manufacturing depends on the demands and profiles of products, technical requirements and capacities. These requirements are constantly changing with the result that manufacturing systems need to be permanently adapted. (Schimmelpfeng 2001), (Westkämper and Niemann 2002)

1.2 Goals of a sustainable product life cycle management

The new paradigm of optimising a technical product's cost-benefit is orientated not only towards economic but also towards environmental aspects by applying ecological criteria. It assumes that the concentration on core competencies and specialisation offers new potentials to add value or reduce the cost of usage by industrialising services and disassembly.

A common understanding between manufacturers and users is a prerequisite for activating potentials in order to obtain the maximum benefit from each technical product during its life cycle and to fulfil economic and environmental objectives (Figure 1.5).

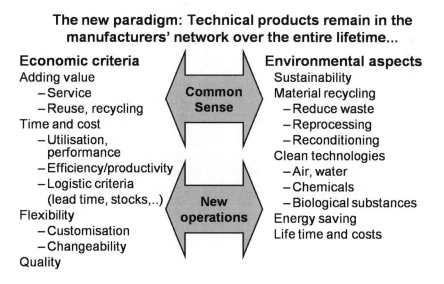

The new paradigm: Technical products remain in the manufacturers' network over the entire lifetime...

Economic criteria
Adding value
 – Service
 – Reuse, recycling
Time and cost
 – Utilisation, performance
 – Efficiency/productivity
 – Logistic criteria (lead time, stocks,..)
Flexibility
 – Customisation
 – Changeability
Quality

Common Sense

New operations

Environmental aspects
Sustainability
Material recycling
 – Reduce waste
 – Reprocessing
 – Reconditioning
Clean technologies
 – Air, water
 – Chemicals
 – Biological substances
Energy saving
Life time and costs

Fig. 1.5: Objectives of life cycle management.

Common sense and active optimisation demand technical solutions that link products at any point in the time of their entire life cycle to the information networks of manufacturers and users. This can be achieved by integrating technical products into global IT networks and electronic services. It is evident today that we have the technologies to do this and also to follow the technical trend for developing intelligent machines connected up to communications systems. (Westkämper 2006) Following this new paradigm, the vision is to permanently link products to manufacturers' networks. Communication is the platform for any product-orientated service with the aim of achieving maximum benefit over a product's lifetime.

1.3 Approach of the book

This book takes a holistic approach to discuss the issues and interdependencies presented in the preceding chapters regarding the life cycle of products. Figure 1.6 gives a schematic overview of the structure of the book.

Chapter 1 is concerned with the general industrial and societal conditions which need to be taken into account in the design of sustainable product life cycles. Not only do the industrial requirements of the manufacturer and user of the products play a major role here but also legal and superior societal aspects.

Chapter 2 deals with aspects regarding the modelling of product life cycles. As well as relevant standards, an emphasis is also placed on special concepts for modelling products. New and innovative approaches show how the life of a product can be digitally modelled and analysed to generate sustainable improvements throughout its life cycle.

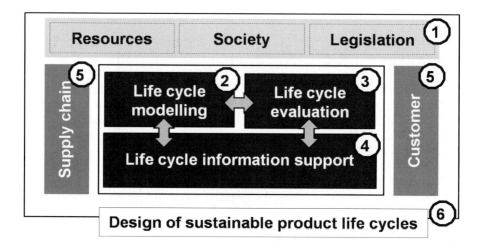

Fig. 1.6: Structure of the book.

With regard to industrial relevance, the consideration of life cycle costs plays a crucial role. Therefore, a method is presented in Chapter 3 for recording and calculating costs over a product's life. The life cycle costing method has proved itself on industrial application and is implemented in many branches of industry. These concepts are described in the book and have been extended by a method for life cycle controlling in order to overcome the problems of long-term planning horizons and associated cost uncertainties.

As far as the sustainable design of products is concerned, uncertainty regarding future developments is a huge problem in itself. No-one can really predict the future. However, the implementation of experience gained as well as the optimal exploitation of existing current usage data helps to constantly optimise product utilisation. Chapter 4 is therefore devoted to life cycle information support. The chapter demonstrates the benefits which can be gained from the consistent acquisition and analysis of data obtained during a product's entire life cycle.

Chapter 5 extends this perspective by integrating additional partners throughout the product life cycle. In the process, suppliers and other knowledge sources are linked to form a customer-orientated network. This network is available for use by customers as a partner to optimise products from a life cycle point of view. In this way, long-term added value partnerships are created which can be measured, organised and cared for using the instrument of customer lifetime value.

Chapter 6 combines the various concepts mentioned in the preceding chapters and uses them to develop a consistent holistic method for the sustainable design of product life cycles. The new and innovative method integrates the different aspects into product life cycle management takes the various needs and requirements of the various stakeholders during the life cycle into account. Not only is product development recorded but also methods and control loops for continuous product optimisation throughout all the phases of life.

1.4 References concerning chapter 1

(Alting et al. 1997) Alting, L., Hausschild, M., Wenzel, H.: Environmental Assessment of Products, Volume 1: Methodology, tools and case studies in product development, Chapman and Hall, London 1997.

(Alting and Jorgensen 1993) Alting, L., Jorgensen, J.: The Life Cycle Concept as a Basis for Sustainable Industrial Production, Annals of the CIRP, 1993, vol. 42 (1): 163-167.

(Alting and Legarth 1995) Alting, L., Legarth, J. B.: Life cycle engineering and design, CIRP keynote paper, Annals of the CIRP, 1995, Vol. 44(2).

(Anderl et al. 1997) Anderl, R., Daum, B., John, H., Pütter, C.: Cooperative product data modeling in life cycle networks; Life Cycle Networks Chapman & Hall London, Weinheim New York Tokio Melbourne Madras 1997.

(Blanchard 1978)   Blanchard, B.: Design and manage to life cycle cost, Portland, Or.: M/A Pr., 1978.

(Brussel and Valckenaers 1999) Brussel, H. van, Valckenaers, P. (Hrsg.): Katholieke Universiteit <Leuven> / Department of Mechanical Engineering / Production Engineering Machine Design Automation (PMA): Intelligent Manufacturing Systems 1999: Proceedings of the Second International Workshop on Intelligent Manufacturing Systems, September 22-24, 1999, Leuven, Belgium.

(Feldmann 2002) Feldmann, K.: Integrated Product Policy - Chance and Challenge: 9th CIRP International Seminar on the Life-Cycle Engineering. April 09.-10., Erlangen, Germany. Meisenbach, Bamberg, 2002

(Fullana and Puig 1997) Fullana, P., Puig, R.: Análisis del ciclo de vida. Rubes Editorial S.L., Barcelona 1997.

(Harjula et al. 1996)Harjula, T., Rapoza, B., Knight, W.A., Boothroyd, G.: Design for disassembly and the environment, Annals of the CIRP, 1996, Vol. 45(1): 109-114.

(Hoeck 2005)Hoeck, Hendrik: Produktlebenszyklusorientierte Planung und Kontrolle industrieller Dienstleistungen im Maschinenbau. Aachen : Shaker, 2005 (Schriftenreihe Rationalisierung und Humanisierung 76). Aachen, RWTH, Fak. für Maschinenwesen, Diss. 2005

(Kondoh et al. 1998) Kondoh, S., Umeda, Y., Yoshikawa, H.: Development of upgradable cellular machines for environmentally conscious products, Annals of the CIRP, 1998, Vol. 47(1): 381-394.

(Niemann 2003) Niemann, J.: Life Cycle Management, In: Bullinger, H.-J. (Hrsg.), Warnecke, H. J. (Hrsg.), Westkämper E. (Hrsg.), Neue Organisationsformen im Unternehmen - Ein Handbuch für das moderne Management, 2. neu bearbeitete und erweiterte Auflage, Berlin u. a.: Springer Verlag 2003.

(Niemann and Westkämper 2004) Niemann, J., Westkämper, E.: Life cycle product support in the digital age. In: CIRP u.a.: Design in the Global Village/CD-ROM: 14th International CIRP Design Seminar, ElMaraghy, W. (Chair), May 16.-18., Cairo, Egypt, Windsor, Ontario, CA, 2004

(Qu-Yang and Pei 1999) Ou-Yang, C., Pei, H.N.: "Developing a STEP-based integration environment to evaluate the impact of an engineering change on MRP", International Journal of Advanced Manufacturing Technology, 15, pp. 769–779, 1999.

(Rifkin 2000) Rifkin, J.: Access, das Verschwinden des Eigentums, 2. Auflage, Campus Verlag, Frankfurt/New York, 2000.

(Schimmelpfeng 2001) Schimmelpfeng, K.: Lebenszyklusorientiertes Produktionssystem-controlling, Deutscher Universitäts-Verlag GmbH, Wiesbaden 2002, zugl. Habilitationsschrift, Universität Hannover, 2001.

(Steinhilper 1998) Steinhilper, R.: Remanufacturing – The Ultimate Form of Recycling, Stuttgart: Fraunhofer IRB 1998.

(Sivard 2003) Sivard, G.: An integrated architecture for functional products, Proc. of ICED 2003, International Conference on Engineering Design, Stockholm, 19-21 August, 2003

(Seliger 1997) Seliger, G.: More use with fewer resources – a contribution towards sustainable development; Life Cycle Networks Chapman & Hall London, Weinheim New York Tokio Melbourne Madras 1997.

(Sheng et al. 1998) Sheng, P., Bennet, D., Thurwachter, S.: Environmental-based systems for planning of machining, Annals of the CIRP, 1998, Vol. 47(1): 409-414.

(Westkämper 2000) Westkämper, E.: Technical Intelligence for Manufacturing. Third World Congress on Intelligent Manufacturing Processes and Systems. Cambridge, MA ; 2000

(Westkämper 2006) Westkämper, E.: Einführung in die Organisation der Produktion. Strategien der Produktion. Berlin; Heidelberg: Springer, 2006

(Westkämper et al. 2000) Westkämper, E., Alting, L., Arndt, G.: Life Cycle Management and Assessment: Approaches and Visions Towards Sustainable Manufacturing. In: Annals of the CIRP, Manufacturing Technology 49, 2, pp. 501-522., 2000

(Westkämper and Niemann 2002) Westkämper, E., Niemann, J.: Life Cycle Controlling for Manufacturing systems in web-based environments, In: CIRP u.a.: CIRP Design Seminar: Proceedings, 16-18 May, Hong Kong 2002.

(Zülch et al. 1997) Zülch, G., Schiller, E.F., Müller, R.: A disassembly information system; Life Cycle Networks. Chapman & Hall, London, Weinheim, New York, Tokio, Melbourne, Madras, 1997

2 Life cycle modelling

In order to manage life cycle objects and for actors involved in the life cycle to collaborate efficiently, life cycle objects should first be modelled. This approach forms a basis for implementing product life cycle management (see Figure 2.1).

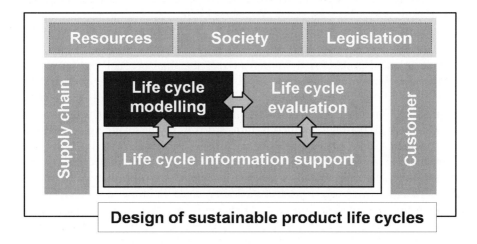

Fig. 2.1: Structure of chapter 2.

To model product life cycle objects during the product life cycle, a research project called PROMISE was carried out. The PROMISE project (FP6 507100 and IMS 01008) was launched by the European Union to develop a new generation of product information tracking and flow management systems. For more detailed information, see (Kiritsis 2004), (PROMISE 2004) and (Kiritsis and Rolstadås 2005) or visit the project website: www.promise.no. The report presents the state of the art regarding product and process modelling methods.

Manufacturing competitiveness supports sustained growth and profits by promoting customer loyalty through the creation of high-value products for the dynamic global market (Biren 1996). The demand for higher-quality and lower-cost products with shorter development times has forced industries to focus on new product development strategies. The well known computer integrated manufacturing (CIM) is an advanced manufacturing system which uses information technology and involves the interconnection of various technical and management functions within a company (Harrington 1973). Concurrent engineering (CE) has been proposed and defined by many researchers as a means of minimising product development times (Barkan 1988), (Chan and Gu 1993), (Winner 1988), (Ou-Yang and Pei 1999), (Zhao 1998), (Bhandarkar and Ngai 2000), (Seltes 1978), (Hunag and Mak 1999).

CE is a systematic approach to the integrated, concurrent design of products for dynamic global markets (Biren 1996).

Other strategies, such as lean production, (Womack et al. 1989), (Clark and Fujimoto 1991), agile manufacturing (Nagal and Dove 1991), (Youssef 1992) virtual manufacturing (Onosato and Iwata 1993), (Iwata et al. 1995), holonic manufacturing (Winkler and Mey 1994), Matheus 1995), continuous acquisition and life cycle support (CALS – formerly known as computer-aided acquisition and logistics support) (Baumann 1990) and knowledge-based intelligent system approach (Aldalondo et al. 2000), all contribute in different ways towards product development from conceptual design to production and distribution in order to enhance industrial competitiveness.

Represented by all the above-named strategies, the integrated product design and development process is the foundation of the final product realisation and involves numerous management and information technologies. However, technology only makes these strategies possible and does not actually create them (Rasmus 1993). Changes in market conditions are driving the implementation of new emerging technologies, which in turn is driven by the changing processes it has to support.

Product modelling is considered to be one of the key technologies that enables the realisation of these strategies during product development activities. Figure 2.2 shows the scope of complete product modelling made up of the four major factors of enterprise objectives, enterprise development strategies, enterprise manufacturing resources and product realisation processes which are related to the development, maintenance and subsequent usage of product models within an industry.

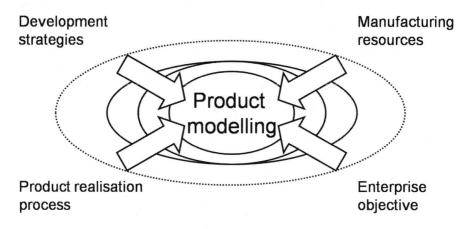

Fig. 2.2: The realm of product modelling.

These four aspects are interlinked with the information flow of product modelling. It is understandable that Krause (Krause and Schlingheider 1993) stated that "the issues of information processing for product modelling are very complex in engineering practice". The product realisation process models the product life cycle from ideas through to final details. There are a number of parallel subproduct models with varying information contents and structures. Hence, product modelling technology is critical (Trappey et al. 1997), (Murphy 1950), (Quaissi et al. 1999).

The purposes of life cycle modelling are three-fold:

- The model should guide *life cycle design* by identifying, in particular, the holistic nature of life cycles. While we often pay too much attention to individual life cycle processes, it is critically important to consider their holistic nature, such as interferences between product life cycle processes, total life cycle costs and also environmental influences, etc.
- The model should be used in life cycle simulation. The model described in this section was indeed developed to serve as a simulation model based on a discrete event simulation technique (Umeda et al 2000).
- The model should serve as a reference in the evaluation of product life cycles, not only for life cycle simulation, but also for LCA, etc. While a life cycle scenario is typically assumed for an LCA case study, it is generally an average case. However, extreme cases also exist on a number of occasions, especially for consumer products.

The following example shows a life cycle model for evaluating the life cycle design of a product using life cycle simulation. A life cycle model consists of a product model and a process network. Life cycle models, each of which representing a life cycle design, contain multiple processes in the process network such as production, operation, maintenance, disassembly and reuse. It is assumed that a product model is modular in order to reduce costs associated with maintenance, disassembly and reuse. The modular structure of the product of each life cycle model is optimised with respect to multi-objective criteria (e.g., amount of waste, material and energy consumption, corporate profits). Umeda et al. (2000) presents five different types of life cycle model for evaluating life cycle designs which are either commonly used nowadays or potentially used as sustainable alternatives.

- Conventional type: in this life cycle, a customer buys a product and throws the entire product away without having it repaired or maintained when the product becomes faulty or obsolete. This type of life cycle is commonly used for consumer electronics.
- Recycling type: this is similar to the conventional type with the exception that discarded products are recycled. For the sake of simplicity, it is assumed that metals are material-recycled and plastics energy-recycled, reflecting real world practices. For example, automobiles currently belong to this life cycle type.
- Reuse type: in this life cycle, modules from collected discarded products are reused or recycled. This decision is made based on the elapsed lifetime of the products concerned. Printers are one examples of this type.
- Maintenance type: here, the customer continues to use the product after purchase by maintaining it (i.e., replacing defective modules with new modules) and the defective modules become waste. However, if maintenance becomes

too expensive for a customer, the old product is replaced with a new one. For example, machine tools belong to this type of life cycle.

- PMPP (Post Mass Production Paradigm) type (Tomiyama 1997), (Umeda et al. 2000): this life cycle consists of maintenance, modular and component reuse as well as recycling processes. These processes are managed accordingly (for example, products are leased rather than sold resulting in a higher rate of collection than with other types). With appropriate life cycle management, this type can become an ideal sustainable life cycle. The above four life cycle models have in fact been derived from the PMPP type model by changing parameter values.

Although these models have been simplified, they are helpful when considering the essential advantages and drawbacks of recycling, reuse and maintenance options. Here, life cycle simulation is useful to quantitatively evaluate the performance of the models at an early stage in product development.

Andreasen (Andreasen 1992) takes a total life cycle view of product modelling by introducing the notion of product life phase systems. During its lifetime, a product interacts with various life-phase systems such as manufacturing systems, distribution systems or destruction systems. These systems are developed and/or configured concurrently with the product and are referred to as life cycle systems (LCS).

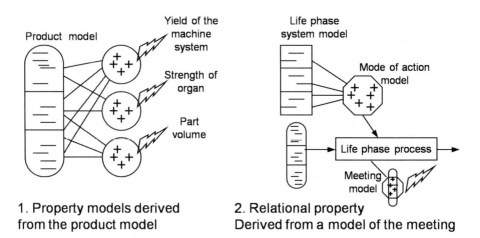

1. Property models derived from the product model

2. Relational property Derived from a model of the meeting

Fig. 2.3: Meeting between a product and a life-phase system.

Mortensen and Andreasen (Mortensen and Andreasen 1996) continue discussions regarding interactions between a product and its life-phase systems by introducing the *meeting theory* where relations between two design objects, synthesised into two separate chromosomes, are called meetings (Figure 2.3). Although it is clear that the product and the LCS are two separate technical systems which - due to high interdependency - need to be developed in harmony with each other, further explanation of the relationship between a product and its LCS is not included in the scope of the original theory of domains.

An important aspect to be considered when applying the engineering design to LCS design is that of transformation processes. A life cycle process could be regarded as being a transformation process which is executed by a transformation system – a life cycle system. In such a process, an operand (product) is transformed from an initial state (e.g. material) into a desired state (e.g. manufactured product).

Product components embody organs which, in their turn, realise functions which carry out the process transforming customer operands. The physical properties of product components are managed in the life cycle processes executed by the LCS.

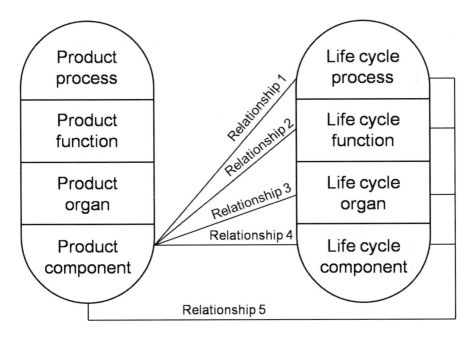

Fig. 2.4 : The relationship between a product and its LCS.

For example, a product's components' physical properties are determined in the manufacturing process executed by the manufacturing resources within a manufacturing system. LCS components embody the LCS organs which realise LCS functions which carry out the life cycle processes. Therefore, it can be suggested that a relationship exists between products and their LCS established during the execution of the development process and expressed in the causality links between the domains of the product system and LCS. (Figure 2.4)

2.1 Process network

A product life cycle is made up of a network of processes. Each process has a specific functionality within a product life cycle such as manufacturing, operation, recycling and remanufacturing. Management of a product life cycle requires understanding of the behaviour of these process and their interdependencies. For example, Figure 2.5 shows important processes within the product life cycle of a modularised product.

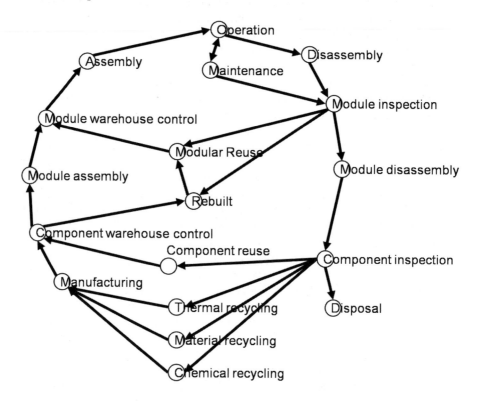

Fig. 2.5: A process network in a product life cycle. (Tomiyama 1997), (Umeda et al. 2000).

The behaviour of a process in the product life cycle depends on the input-output flows of components, modules and products from other processes. Although information can be detailed and quantified to describe these processes such as process parameters and decision criteria (Umeda et al. 2000), the behaviour of each process in the product life cycle can be based on the following description.

- Market: the size of the market is constant or fluctuates dynamically. Only the target product is sold in this market. Once the target product has started to be sold, a consumer can only buy the target product and the market is saturated with the product within a certain period of time. Competition among different products in the same market can be included in the model.

- Operation and maintenance: a customer buys a new product at purchase price. This price does not include maintenance or recycling costs. Each component breaks down randomly throughout its lifetime and the customer repairs the product by replacing a defective module with a functioning one. The maintenance costs paid by the customer pays for each maintenance task are formulated as being the sum of the module price and maintenance fee. However, if the cost of maintenance is too high or the product breaks down too often, the customer replaces the entire product with a new product. This customer's judgement is based on customer maintenance preference and the minimum acceptable Mean Time Between Failures (MTBF).

- Collection and product disassembly: discarded products are collected for the product disassembly process and the rest are simply disposed of. This judgement is parameterised and indicates that the collection system of discarded products is not ideal. The product disassembly process strips collected products into modules and sends them to the module inspection process.

- Module inspection: modules from disassembled discarded products as well as those collected from the maintenance process are sent to the module inspection process. The module inspection process classifies collected modules into the following groups: (1) the module is not defective and can thus be reused, (2) the module contains non-reusable components. In the first case, the module is sent to the module warehouse for light reconditioning (e.g., cleaning). This is determined by evaluating all components in the module using a reusability criterion. The criterion is described as a function of nominal lifetime, operation time, guarantee period and reusability of component. In the second case, the module goes either to the rebuilding process or the module disassembly process. This judgement is formalised as a function of the proportion of reusable components in the module; therefore, if it is easy to recover the module, it goes to the module rebuilding process but if the module contains many non-reusable components, it goes to the module disassembly process instead.

- Rebuilding: in the module rebuilding process, modules are remanufactured by replacing non-reusable components with working ones that came from the component warehouse. Rebuilt modules are sent to the module warehouse.

- Module disassembly and component inspection: modules are disassembled into components which are then inspected in the component inspection process. Thus, if a component satisfies the reusability criterion, it is sent to the component warehouse after light reconditioning. Otherwise, the component goes to the recycling process if it is recyclable and, if not, it is simply disposed of. The recyclability of components is specified by the user in the product model.

- Recycling: in the recycling process, components are recycled either into material or energy. If a component is made of metal it is recycled into material; if made of plastic, it is recycled into energy and other materials are simply disposed of. In the material recycling process, a certain weight ratio of input components is produced as recycled materials which will be used for component manufacturing and the rest is disposed of. Although the energy recycling process generates energy for other processes, it also generates waste, i.e. a certain weight ratio of input plastic components becomes waste. This ratio depends on the properties of the components concerned.

- Warehouses: modules in the module warehouse are used for maintenance as spare parts and in the assembly of new products. If the number of modules in the warehouse is below a certain threshold, new modules are manufactured. The component warehouse works in the same manner.

Such a representation of the product life cycle makes it possible to explicitly treat all product life cycle processes and parameters as design parameters of the life cycle of products and to view their interdependencies from a systematic perspective.For process modelling, there are many modelling languages available such as PSL, XPDL and BPML. For specific modelling, parts of UML, EPC or OPM can be used. The details are described in the following.

2.1.1 PSL (Process Specification Language)

The Process Specification Language (PSL) (Version 1.0) which was developed at the National Institute of Standards and Technology not only identifies and formally defines but also structures the semantic concepts intrinsic to the capture and exchange of discrete manufacturing process information (http://www.mel.nist.gov/psl/). Process data are used throughout the life cycle of a product, from early indications of manufacturing processes pinpointed during design, process planning and validation right up to production scheduling and control. Additionally, the notion of process also underlies the entire manufacturing cycle, coordinating workflows within engineering and shop floor manufacturing.

2.1.2 XPDL (XML Process Definition)

The XML Process Definition Language (XPDL) is a workflow process definition meta-data model to provide a common method for accessing and describing workflow definitions. This meta-data model identifies entities commonly used within a process definition. A variety of attributes describe the characteristics of this limited set of entities. Based on this model, vendor specific tools can transfer models using a common exchange format (WfMC 2002). A key element of XPDL is its ability to be extended to handle information used by a variety of different tools. However, XPDL will never be capable of supporting all additional information requirements in all tools. XPDL supports a number of different approaches based upon a limited number of entities describing a workflow process definition (the *Minimum Meta Model*). One of the most important elements of XPDL is a generic construct which supports vendor specific attributes for use within the common representation.

2.1.3 BPML (Business Process Modelling Language)

The Business Process Modelling Language (BPML) is a meta-language for modelling business processes in the same way that XML is a meta-language for modelling business data. BPML provides an abstracted execution model for collaborative and transactional business processes based on the concept of a transactional finite-state machine (Arkin 2002). BPML defines activities of varying complexity, transactions and compensation, data management, concurrency, exception handling and operational semantics. BPML provides a grammar in the form of an XML schema to enable the persistence and interchange of definitions across heterogeneous systems and modelling tools. BPML can be used to define the detailed business process behind each service. BPML maps business activities to message exchanges. It can be used for the definition of enterprise business processes, the definition of complex Web services and multiparty collaborations.

2.1.4 UML (Unified Modelling Language)

Among Unified Modelling Language (UML) diagrams, the sequence diagram, swimlane chart and state diagram can be used for process modelling from the object-orientated viewpoint.

Sequence diagrams describe inter-object behaviour, i.e. the way in which single objects interact through message passing in order to fulfil a task. They are also called interaction diagrams. A sequence diagram is mainly used for modelling a scenario to show the flow of an operation or case of use. The diagrams can be extended to describe entire algorithms (there are symbols for distinguishing between cases and repetition) but often lose their clarity if used in this way.

Swimlane charts are often utilised together with activity diagrams where different parts of an organisation or a large system are divided into different swimlanes. A swimlane is a method for grouping activities performed by the same actor of an activity diagram or grouping activities in a single thread. Swimlanes are regions in a diagram which contain active objects and which are separated by vertical or horizontal lines.

State chart diagrams describe intra-object behaviour, i.e. the possible consequences of states which an object of a class may go through during its life, either from its creation till its destruction or during the execution of an operation. State chart diagrams are based on general finite automatons.

2.1.5 EPC (Event Process Chain)

In the years from 1990 to 1992, the foundational conceptual work for an SAP reference model was conducted by the SAP AG and the IDS Scheer AG in a collaborative research project. The outcome of this project was the Event Process Chain (EPC) which has been used for designing reference process models in SAP.

EPCs also became the core modelling language in the Architecture of Integrated Information Systems (ARIS). EPCs are directed graphs consisting of events, functions and connectors to visualise the control flow. Each EPC starts with at least one event which triggers a function and in turn leads to a new event. All functions and events are connected by control arcs (PROMISE 2005). (Scheer 1998a), (Scheer 1998b)

2.1.6 OPM (Object Process Methodology)

Object Process Methodology (OPM) is an approach for modelling a real or conceptual world with the help of an object-orientated approach. OPM takes a fresh look at modelling complex systems comprising of humans, physical objects and information. OPM is also a formal paradigm for systems development, life cycle support and evolution. It can also be used to support the structuring of people's intuition and training of thought.

OPM supports not only language but also graphical notation – OPD (Object-Process Diagram), making it very helpful for conceptualising people's thoughts and communication with each other. Graphical notation can be translated into OPL (Object-Process Language) which represents the model. Two CASE tools are available for supporting OPD and OPL: OpCat and Systematica.

As OOA/D is a well-known methodology, anyone can easily understand OPM by making a comparison between two methodologies. OOA/D is generated in the software engineering domain while OPM is a top-down representation of a system without the constraints of a programming language.

2.2 Framework for a networked life cycle management

The whole product life cycle includes all phases from product generation and usage right up to product disposal. It consists of the following life cycle activities: design (conceptual design and detail design), production (procurement, manufacturing and assembly), logistics (distribution), usage, maintenance (service), collection, remanufacturing (disassembly, refurbishment, reassembly, etc.), reuse, recycling and disposal (Hong-Bae et al. 2005a). Figure 2.6 shows the conceptual model of the whole product life cycle.

The product life cycle generally has complicated flows of activities and information. Therefore, it is important to describe them in order to make it possible to control and steer the process and information flows of the product life cycle. The description can be used a basis for process and data integration. For this purpose, modelling issues require consideration.

First of all, the necessity of a life cycle modelling framework needs to be described, followed by the modelling issues of product life cycle data.

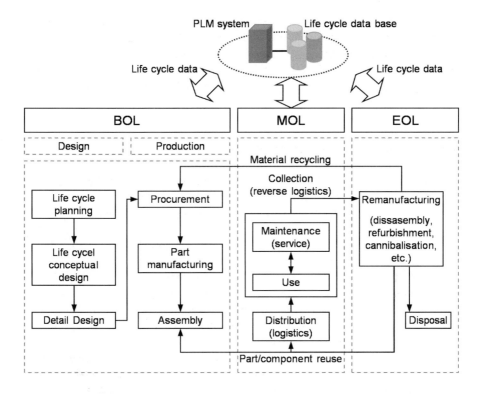

Fig. 2.6: Whole product life cycle.

2.2.1 Defining the product life cycle modelling framework

To enhance the performance of life cycle operations over the entire product life cycle, all life cycle objects performed along the product life cycle have to be designed and coordinated and their efficiency managed. The numerous modelling works in existence can be classified into two categories: enterprise modelling methodologies and product life cycle modelling, as shown in Table 2.1.

Although considerable research work concerned with enterprise or product life cycle modelling has been carried out in the past, up till now little attention has been paid to the whole life cycle modelling framework (Nonomura et al. 1999), (Kimura 2002). Although CIMOSA and PERA consider the life cycle concept, they focus on the development of manufacturing systems or enterprise business. Moreover, previous product life cycle modelling methods lack integrated views throughout the enterprise life cycle (Tipnis 1995), (Kimura and Suzuki 1995). For example, Ming and Lu (Ming and Lu 2003) addressed the architecture for PLM but only focused on the process viewpoint. As a result, enterprise modelling frameworks were too sizeable to describe PLM models and the previous life cycle modelling research was too conceptual and implicit.

Classification	Previous Research
Enterprise modelling	IDEF (Integrated computer aided manufacturing DEFinitions methodology) (Mayer 1994),
	IEM (Integrated Enterprise Modelling) (Vernadat 1996),
	PERA (Purdue Enterprise Reference Architecture) (Vernadat 1996),
	CIMOSA (Open System Architecture for CIM) (Bruno and Agarwal 1997),
	ARIS (Architecture for integrated Information System) (Scheer 1998a, 1998b),
	UEML (Unified Enterprise Modelling Language) (Vernadat 2002)
Product life cycle modelling	IDEF (Integrated computer aided manufacturing DEFinitions methodology) (Mayer 1994),
	IEM (Integrated Enterprise Modelling) (Vernadat 1996),
	PERA (Purdue Enterprise Reference Architecture) (Vernadat 1996),
	CIMOSA (Open System Architecture for CIM) (Bruno and Agarwal 1997),
	ARIS (Architecture for integrated Information System) (Scheer 1998a, 1998b),
	UEML (Unified Enterprise Modelling Language) (Vernadat 2002)
	Conceptual life cycle modelling framework with IDEF (Tipnis 1995)
	Conceptual architecture and the key components for total product life cycle design supporting system (Kimura and Suzuki 1995)

Table 2.1: Modelling methods.

As a result, it is necessary to develop the life cycle modelling framework to describe the behaviour of whole product life cycle objects: product, process and resource. Within this framework, a definition for representing the product life cycle objects is required. For each life cycle object model modelling constructs and methods need to be defined in a standard and flexible way for adaption to various application domains. In addition to this, the interactions between three life cycle object models (product, process and resource) should also be described, taking the integration of life cycle object models into account.

2.2.2 Modelling product life cycle data

Product life cycle operations contain the planning, execution, control and documentation of all processes throughout the entire product life cycle (Abramovici et al. 1997). This means that the scope of information during a product life cycle is extended beyond the product itself and encompasses all intellectual capital including products, processes and resources (Collaborative visions 2002). Therefore, during the whole product life cycle, a great deal of life cycle information such as CAD drawings, technical documents and structured and unstructured data is created, changed, transferred, stored and converted by and between different people and application systems. This results in complex information flows over the whole life cycle (Figure 2.7).

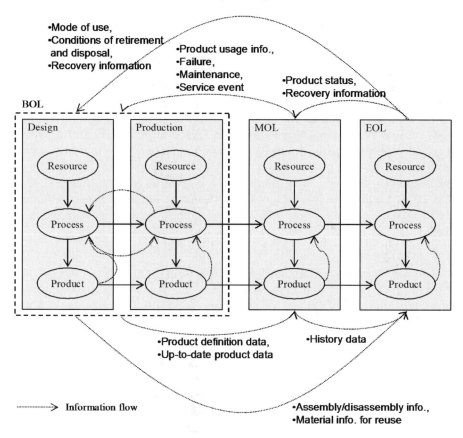

Fig. 2.7: Example of a life cycle information flow.

Efficient management of the product life cycle information is essential and this can play an important role in the analysis and decision-making of various operational issues in the product life cycle.

Up till now, considerable attention has been paid to the information modelling issues of product life cycle data. For example, Scheidt and Zong addressed the structure of life cycle history data to achieve the reusability of electronic modules (Scheidt and Zong 1994). Klausner and Grimm described the data processed by ISPR (Information System for Product Recovery) at a product's EOL (Klausner and Grimm 1998). Goncharenko et al. proposed an approach for gathering and utilising product feedback information over the product life cycle through the use of life cycle design methodologies (Goncharenko et al. 1999).

Furthermore, (Fangyi et al. 2002) proposed a conceptual information model to define and describe product life cycle environment characteristics. (Schneider and Marquardt 2002) defined the product life cycle of chemical processes and described elements of design life cycle: activities, all kinds of information, resources - mainly software tools - and the organisational unit. They also proposed an information modelling method in the chemical design life cycle for capturing the objects and concepts characterising the life cycle together with their relationships and interdependencies. (Kiritsis et al. 2003) defined the data recording structure which allows maintenance personnel to store relevant data obtained during maintenance operations for use in design. Additionally, (UGS PLM 2004) proposed an open standard, called PLM XML, to facilitate high-content product life cycle data sharing over the Internet. On the other hand, there have also been several research works related to information interoperability. For example, (Fenves et al. 2003) proposed a framework for product information interoperability to access, store, serve and reuse whole product life cycle information. (Terzi et al. 2004) reviewed interoperability standards along the product life cycle. In addition to this, (Lubell et al. 2004) described an overview of STEP (Standard for the Exchange of Product Model Data), XML and UML (Unified Modelling Language) for application in PLM.

However, previous research works have some limitations. Despite the importance of product life cycle data, most did not address the subject of representing the bulk of product life cycle data efficiently, especially that of MOL and EOL. Only a few works were concerned with MOL life cycle data. Previous works which focused on EOL life cycle data did not deal with information modelling methods. Moreover, in the past very little attention has been paid to clarifying the structure and semantics of product life cycle data. Many researchers also used XML to describe product life cycle information. XML schemata have the limitation that they do not provide any information about the content, i.e. meta-information (Hong-Bae et al. 2005b).

To overcome these limitations, a modelling method for product life cycle data needs to be developed. In order to trace product life cycle information efficiently, it is necessary to design and manage it in a systematic way. This requires an in-depth understanding of the semantics and structure of product life cycle data over the whole life cycle. First of all, the information relevant to product life cycle data needs to be clarified. It is also necessary to know what product life cycle data is required for each operational issue and each life cycle phase and to classify the information into different types depending on its characteristics. In addition, it is essential to design the data structure which describes the contents of product life cycle data, i.e. meta data and to develop an information modelling framework for managing product life cycle meta data. Finally, methods for retrieving the information and knowledge from life cycle meta data and for analysing it also require discussion.

2.3 Product models

Information exchange in the manufacturing domain is an issue which is becomingly increasingly important to resolve as the manufacture of parts becomes more and more dependent on the use of computerised information systems. Many studies have been performed which show how standardised information models can be utilised to enable information exchange between computer systems. (Al-Timini and MacKrell 1996), (Johanson 2001), (Nielsen 2003)

However, the information models developed for specific areas of manufacturing are too specific. The specific nature of the models means that they are unable to cope with developments in the area. There is also a danger in over-generalisation as the benefit a standardised model is lost if different implementations use it in different ways.

The duality problem can be solved by combining a general model with a reference data library containing the specific concepts within a domain. The specific information items necessary to represent information in a domain are accessible while still enabling the use of a general model. The general model allows for stability over time because the model does not have to be updated as often as a more specific model. It is the specific concepts of domains of applications - ideally represented as an ontology - which have to be maintained.

Many modelling works related to product models have been written and these are described in the following chapters.

2.3.1 Definition of product modelling

In 1950, Murphy made a classic definition of the term *model*: "A model is a device which is so related to a physical system that observations of the model may be used to predict accurately the performance of the physical system in the desired respect" (Murphy 1950). This definition applies mainly to describing the behaviour of physical systems. With the increasing importance of computer-aided technologies, in 1960 Ross introduced the concept of modelling by mathematical means including data, structure, interface and algorithms within the context of CAD/CAM with more relevant behaviours. A model is thus defined as being an abstract specification for domain functions which perform operations.

Modelling simulates the various options in order to make informed decisions early on in the relevant process. It has become the dominant design tool in all aspects of current design. In this work, the term *product* refers to a unit of a function with exact materials, fixed form, designated colour and other features, which is made by an enterprise to satisfy the requirements of a customer. By combining the above definitions of model, modelling and product, the term *product model* can be described as being the sum of all useful information related to a product within the life cycle of its development.

Therefore, product modelling includes all the important processes used to design and develop the product on the basis of product specifications. It has now become the key technology in computer-aided product design and development. Product model data are the result of product modelling action and in different modelling phases, it provides different interrelated model data.

2.3.2 Types of information representation

In the basic derivation, all the information within the product development cycle can be reduced to a modelling process involving different kinds of product models which are interrelated by nature. According to Biren, three types of representational schemes are often employed in this modelling process (Biren 1996):

1. Physical model.
2. Conceptual model.
3. Analytical model.

As shown in Figure 2.8, physical, conceptual and analytical models are used to represent objects (the product) from different points of view and to introduce various aspects of the information. For example, a physical model is useful for conventional physical representation. A conceptual model is more relevant when dealing with the information in the conceptual design phase during product realisation, while an analytical model is more useful for supporting conventional CAD/CAM applications by using parametric or solid modelling.

In this book, the term *product model*, which is used as a mechanism for representing the valid combinations of product information more efficiently, belongs to the category of analytical models. Central to the success of product modelling is the fusion of computer-aided techniques with the designer's knowledge.

Krause and Schlingheider summarise the development of product modelling and the proposed categories of product models, as presented in Figure 2.9 (Krause and Schlingheider 1993). There are five types of product models:

1. Structure-orientated product models
2. Geometry-orientated product models
3. Feature-orientated product models
4. Knowledge-based product models
5. Integrated product models

The structure-orientated product model is the first actual application of a computer-supported product modelling technique in the representation of product structures.

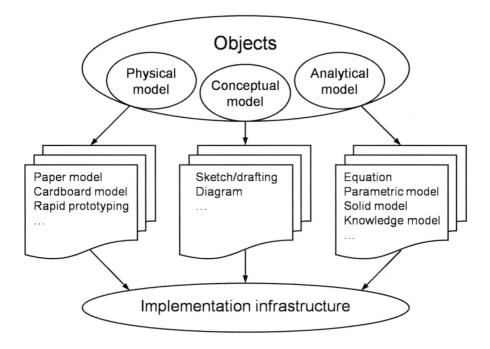

Fig. 2.8: Models representing various types of information.

As the product structure is the core of development activities, such things as specific product data and formats can be stored within this structure-orientated model. Although this kind of model has many limitations with regard to product representation, such as the lack of representation of product shape, it is important in that it provides a basis for further enhancement by other modelling techniques.

The geometry-orientated model was developed as an extension of the structure-orientated model and contains such functions as the representation of product shape including wire frame, surface, solid and hybrid models.

The geometry-orientated model has been widely used to support CAD/CAM and CNC programming applications. It satisfies the requirements of the computer-based representation of the shape of a specific product but is unable to describe non-geometric product information.

The concept of features, usually a form feature, was first put forward for the purpose of representing the general shape patterns of the surface and form of a product as coherent geometric items (Seltes 1978).

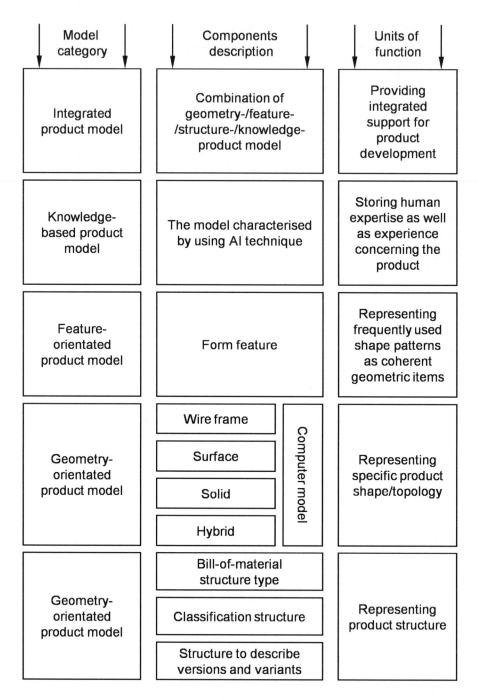

Fig. 2.9: Descriptive summary of product models.

With the subsequent wide use of feature techniques in CAD applications, a feature has become a general information mode for representing a part of a product (Bhandarkar and Ngai 1978).

In the process of product modelling, features can be classified as design features, machining features, assembly features and also abstract features. Each feature has its specific domain of implementation, Figure 2.10.

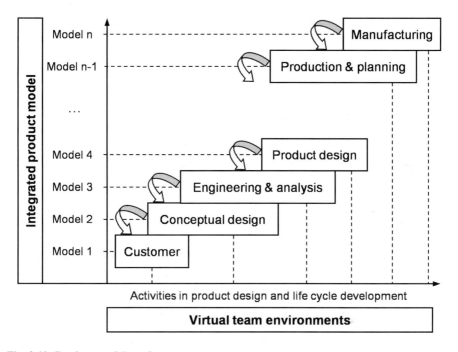

Fig. 2.10: Product models and geometry.

A knowledge-based model is an advanced model which adopts AI (artificial intelligence) techniques. This model supports information reasoning by referring to former designs, human expertise and past experience about a class of products stored in the internal model during the product modelling process.

At present, some implementation methods can be used in the knowledge-based model, such as rule-based reasoning, constraint-based reasoning and object-orientated techniques. The introduction of knowledge in the product modelling domain denotes great progress because the degree of automated reasoning is still an important research topic.

The integrated product model - or global product model - is the functional combination of all the product models discussed above, including structure-, geometry-, feature- and knowledge-based models. The integrated model is used to support all product development activities from the product requirement analysis, conceptual design, detail design, process planning, CNC programming, machining and assembly right up to quality assessment. It can be structured into interrelated multi-view logical models, such as design models and a machining model. Product modelling was also integrated into one of the systematic methods.

The integration of CAD/CAM applications based on the shareable common product information model, including the functions of product data and workflow management, is considered to be one of the key links in the implementation of CE (Qu-Yang and Pei 1999), Zhao 1998).

2.3.3 Existing standards

However, the integration is not fully implemented due to the lack of a unified, single and complete representation of the product and process information model. Consequently, the transfer of product design information from one system to the other often fails because of incompatible or incomplete data (Chan and Gu 1993). Therefore, it is necessary to build an integrated product model for supporting the various activities during the product development cycle, i.e. realising the sharing and exchange of product information within the computer-integrated environment of the enterprise concerned.

2.3.3.1 STEP (STandard for the Exchange of Product)

The STEP (ISO 10303)-Product Data Representation and Exchange standardisation initiative covers the computer-interpretable representation and exchange of product data. STEP is a synonym for all aspects of the international project to develop the technology of product data representation as well as the methodology for creating information model standards and the standards themselves. The method proposed in ISO TC184/SC4 for the ISO 10303 (STEP) suite of standards is to develop a domain-specific model which uses the terminology of that domain and to subsequently map it to a more generic model (ISO 1998). It is becoming more common to create the initial model in a more generic way (ISO 1999), (ISO 2000), (ISO 2005a), (ISO 2005b). This move has been prompted by the realisation that it is an information requirement to represent information generically. Thus, information requirements for the standard go beyond the concepts in a domain.

In the area of product life cycle data where interaction between several different actors is inevitable, the need for a standard is even greater than in most other cases. A standard is already in existence which supports the life cycle aspect of product data: ISO 10303-239, Industrial Automation Systems and Integration – Product Data Representation and Exchange – Part 239: Application Protocol: Product Life Cycle Support, commonly referred to as PLCS (ISO 2005a).

PLCS provides a generic information model which supports the breakdown of the structures of products, processes, requirements etc. with additional information about properties, states, life cycle stages and much more. Together with a reference data library which provides specific concepts such as work order, PLCS forms a solid foundation on which product life cycle data management can be built, based on standardised information representation. The fact that it is a standard will enable different systems from different vendors to be integrated into a common information base for product life cycle data.

The objective of STEP is to provide a means of describing product data throughout the life cycle of a product which is independent from any particular computer system. The nature of this description makes it suitable not only for neutral file exchange, but also as a basis for implementing product databases and archiving data. In practice, the standard is implemented in computer software associated with specific engineering applications, making its use and function transparent to the user. The descriptions are information models which capture the semantics of an industrial requirement and provide standardised structures within which data values can be understood by a computer implementation.

2.3.3.2 PDML (Product Data Markup Language)

Product Data Markup Language (PDML) is an Extensible Markup Language (XML) vocabulary designed to support the interchange of product information between commercial systems (such as PDM systems) or government systems (such as JEDMICS). PDML is being developed as part of the Product Data Interoperability (PDI) project under the sponsorship of the Joint Electronic Commerce Program Office (JECPO) and is supported by several other Federal Government agencies and commercial entities. Three major PDM vendors are active participants in the PDI programme who are developing prototype implementations of PDML. The initial focus of PDML development is legacy product data systems which support the operation of the Defense Logistics Agency (DLA).

PDML is a suite of domain-specific vocabularies integrated into a single, abstract vocabulary. The vocabularies are related via mapping while the specification and translation between the vocabularies is accomplished via the PDML toolkit.

2.3.3.3 Product condition model

Life cycle orientated product design as well as the planning and operation of ecologically and economically optimised product life cycles require the phase-spanning communication and cooperation between all participants in the product life cycle. An increase of the life cycle productivity of resources cannot be achieved through isolated applications, but only through continuous data exchange between IT-systems supporting such fields as design, maintenance, disassembly planning and disassembly.

It is therefore necessary to overcome the heterogeneity of these systems on the one hand and the spatial and chronological differences of the life cycle phases on the other (Kind 2000). In order to achieve this, a data model must be created.

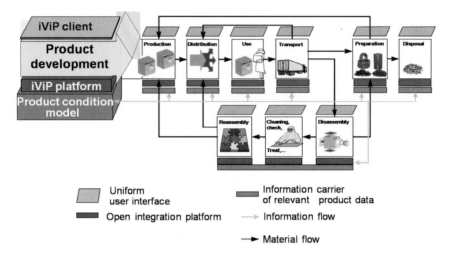

Fig. 2.11: Product life cycle supported by the product condition model.

By defining a model which is able to acquire and represent product condition information in such a way so it can be processed throughout the product life cycle, an information carrier which reduces interfaces and merges systems can be realised (see Figure 2.11) (Krause 2004).

2.4 Applications using product condition data

The application potentials of gathering and supplying condition data in the product life cycle can be categorised into individual product and cluster-based cases. For individual products, maintenance and adaptation processes can be planned or a disassembly simulation carried out. By clustering similar products to form large lots, it is possible to optimise transport routes for maintenance and return, as well as the utilisation planning of disassembly facilities. Furthermore, knowledge about clustered products can be applied in the early adaptation planning stage of the systems in disassembly factories. The provision of product condition information also becomes more important against the background of the modularisation of disassembly systems and factories because its application leads to the fulfilment of requirements regarding module flexibility. (Krause, F.-L. 2004)

With regard to evolving concepts for selling product use instead of the products themselves, clusters of the same product type can support market-orientated adaptation planning and the selection of modules and assemblies for product configuration during re-assembly. The combined consideration of all condition models allows a holistic assessment of ecological and economical parameters. Finally, a data model as a connector linking the product to the integrated IT platform provides a channel for the feedback of information concerning maintenance, use and disassembly to product development.

2.4.1 Conception of the product condition model

An existing physical product manufactured according to the product model has features which deviate from the nominal values as well as features which are not regarded at all in the digital model.

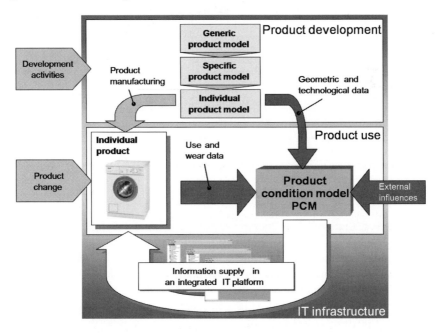

Fig. 2.12: Product condition model for the representation of product changes.

One function of the product condition model (PCM) is to represent these deviations and modifications in a computerised model (see Figure 2.12). The PCM essentially consists of geometric and technological data from the product development and incorporates product- and time-related use and wear information as well as external changes made to product, for example caused by adaptation processes. It can thus be seen as being an extension of the product model beyond the product creation process (see Figure 2.13). (Krause 2004)

The application potentials described above can be classified into online and offline applications. Online application means the acquisition of condition data of a specified product and its use for the operation of the life cycle of the same product, for example initiating a preventive maintenance action or making an end-of-life-decision. In the case of offline application, information from the past is analysed in order to gain knowledge for implementation in the life cycle of new products.

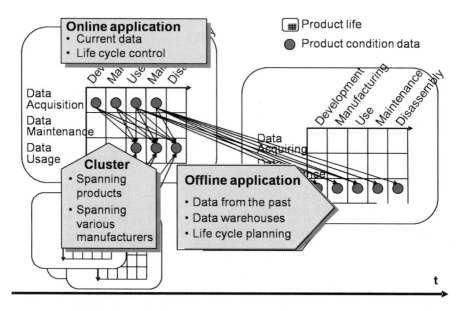

Fig. 2.13: Application principles for condition data

Since the attributes of a product in use change continuously, the PCM has to be adapted dynamically. This requires an information flow from the physical product to the model. The matching between the real product and the digital model is referred to as synchronisation. Synchronisation is carried out within certain intervals, whose frequency and regularity depend on the planned application. The information technology link has to be configured in order to fulfil the respective synchronisation needs. Between two synchronisation points, the divergence of 'real' and 'virtual' conditions increases over time. During synchronisation, the divergence is reduced to the extent the PCM has been designed for. The acceptable tolerance essentially complies with the requirements of the application. Therefore, the PCM exclusively represents product attributes which are relevant either for online or offline applications. As application requirements may change during the life cycle, the model must be both flexible and reactive. Flexibility means the ability of a PCM to represent a large spectrum of possible attributes and reactivity denotes the ability to change as a result of unplanned application needs.

2.4.2 Class structure of the condition model

In order to be able to assign condition data to components of a product, attributes are linked to product structure elements. These elements are modelled in different levels of abstraction to enable clusters to be formed by generalising objects of the product structure. The specification levels distinguished here are the generic level, specific level and the individual level. For example, the generic level may correspond to a product concept. The specific level can be regarded as being a specification on whose basis the production planning is carried out. The individual structure represents the materially-existing product and is assigned using serial or charge numbers.

Each attribute has a predetermined parameter value type. With regard to further STEP processing, these are basically EXPRESS data types. Attributes may relate to direct or indirect condition information. While indirect attributes record influences on a product, such as the rotation count of a bearing, direct attributes express the impact, for example information on the volume of wear. Utilising appropriate transformation methods, indirect attributes can be transferred to direct ones (Martini 2000). A smaller amount of indirect information is needed in order to derivate a multitude of direct condition attributes, making indirect attributes more efficient in terms of acquisition and storage effort. Several research projects focus on the necessary detailed knowledge about product behaviour and physical coherences (Gruzien 2002), (Buchholz and Franke 2003) The transformation of use information into direct condition attributes also allows product conditions to be predicted by projecting the current use mode into the future. (Figure 2.14)

2.4.3 Shifting viewing levels

The attributes assigned to the product components can be processed using miscellaneous aggregation methods. Aggregation in this context means summarising a range of single values. A cluster is formed using a horizontal aggregation method which spans various products. Here, an amount of values of the same attribute but different products is merged. It is also possible to aggregate time series of attributes in a statistical way corresponding to industrial measurement value logging. Vertical aggregation means processing the attribute values of sub-components within a product structure to reach a conclusion regarding the superordinate component. For example, the weight of an assembly equates to the sum of weights of the individual parts. An analysis of the presence of materials requiring separate disposal could also be realised with a vertical aggregation method. The use of aggregation methods makes it possible to transfer information about the condition of an individual product to specific and generic levels. On the other hand, attributes on higher levels can be regarded as being part of the condition description of lower levels. The methods are embedded in the model and remain translucent for the accessing systems. A user is able to access data at his level of viewing, although it may have been generated on another level.

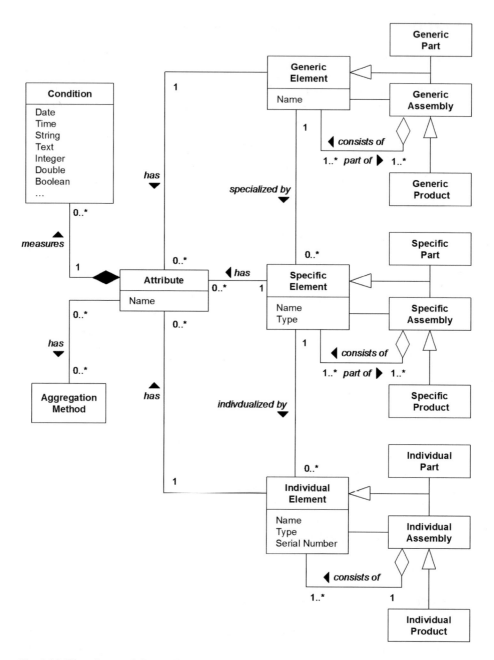

Fig. 2.14: Base classes of the product condition model.

2.4.4 Implementation of the product condition model

A Life Cycle Unit (LCU) is used to acquire condition data directly from the product and to pass them on to the platform. The LCU therefore consists of sensors and a memory chip which buffers the data between the synchronisation intervals. (Gruzien 2002), (Buchholz and Franke 2003) (Figure 2.15)

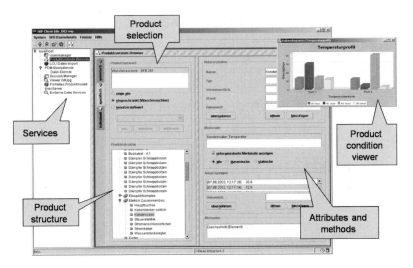

Fig. 2.15: Product condition browser.

A product structure browser enables product condition data to be accessed from the different perspectives. Products which are or have been in use can be selected and navigated in their element structure on the generic, specific or individual level. Closely linked with the browser is the user manager for defining roles. This way suppliers and individual users have access rights and configurations corresponding to their roles. The responsibility for a specific product spectrum with respect to product type, manufacturer and period, viewpoint level as well as access permission to methods and services all need to be determined for the users of the system.

2.4.5 Product model for manufacturing

Today the manufacturing process is a complex phenomenon which requires knowledge of different fields of science such as mechanics, management, economics, etc. In order to build a model, the phenomenon to be modelled should be specified (Figure 2.16, 2.17). The principal tools for simulating the manufacturing process include:

- Installation of a Product Model (PM) for manufacturing,
- Validation of the model by trade experts.

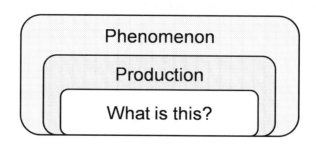

Fig. 2.16: The phenomenon of production.

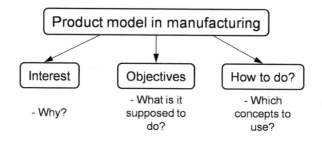

Fig. 2.17: Why the product model for manufacturing?

As well as containing information about its manufacture, the PM also has to be suitable for piloting the manufacture of the product. (Mitrouchev et al. 1998), (Brissaud and Tichkiewitch 2001)

Thus the goals of the model are to:
- Attach descriptive parameters to each product,
- Attach to each product the treatments which it will undergo,
- Launch production in a rational way,
- Trace the product in the course of manufacture (represented by a token),
- Bind each product to its destination (until its delivery),
- Represent the product by a batch of parts (problem of batch bursting),
- Take into account any incidents which may occur,
- Gather information about the quality follow-up,
- Design new products with standards ranges.

The parameters of manufacturing are defined in a similar way as the parameters of design. Both are used later in the Product Model as shown in Table 2.2 below:

Product Model	
In Design	In Manufacturing
Parameters of Design	Parameters of Manufact.
Function (of use) / Need	–
Structure	–
Relations with partners	Relations with partners
Parameters (of Manufact.)	Parameters (of Manufact.)
Quality of product	Quality of product
Control	Control
Ranges (of machining)	Ranges (of machining)
Operations	Operations
Situations	–
Appointment	–
Process	–
Tools	Tools
State	–
Performances	–
Operational time	Operational time
Tool changing time	Tool changing time
History	–
Material	Material
Family of product	Family of product
Geometry	Geometry
Quantity	Quantity
Life cycle / Recycling	–
Times of launching	–
Environment	Environment

Table 2.2: Parameters of manufacturing.

Four hierarchical levels of definition of product parameters are then specified, as summarised in the following table:

Hierarchical levels (by priorities)	
Level 1 (highest)	Life cycle
	Cost
Level 2	Function/ Need
	Geometry
	Structure
	Ranges
	Quality of product
	Control
	Recycling
	Environment
Level 3	Family of product /sub- families
	Material / Characteristics
	State
	Performance
	Appointment
	History
	Situations
	Relations with partners
	Process
Level 4 (lowest)	Resources tools by operat.
	Parameters (of Manufact.)
	Operations
	Quantity

Table 2.3: Levels of product parameters.

The product is defined as being the object which will be transformed. The Product Model is considered both as the *tool* and as the *actor* of manufacturing. Being the link between design and manufacturing, it is intended to accompany the product over its whole life cycle. In fact, the aim is not to await the final description of the product before making information accessible to all the actors of the manufacturing. Consequently, the different aspects of the Product Model are developed which relate to the advance of the product among the various stations of manufacture and the follow-up of the course of operation. This raises the question: what knowledge is required about the product for its manufacture?

- operations of the product range (in which order),
- assignment of the operations,
- sub-operations,
- problem of machine non-availability

The manufacturing knowledge and know-how are distributed among several trade associations. The entities in manufacturing allow each user to work and to express himself in his own language (Brissaud and Tichkiewitch 2001). The essential question is how to achieve this.

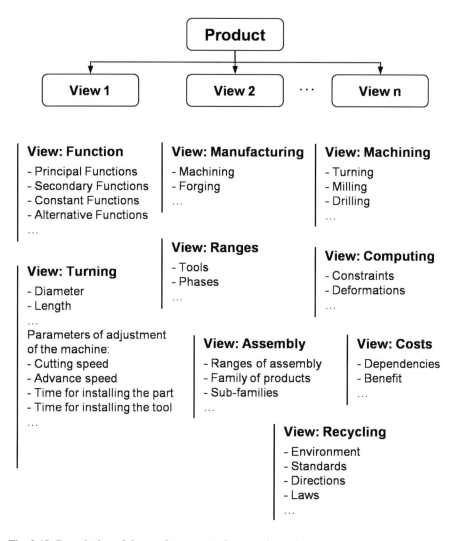

Fig. 2.18: Description of the product as seen from various views.

The following are recommended:
- To build a graph around all the references of production
- To describe the part (object) from the perspective of the manufacturer.

A dictionary of manufacturing rules is compiled to help manufacturers by consolidating their knowledge of manufacturing techniques. A description of the product by views is recommended. The multi-view model for integrated design (Tichkiewitch et al. 1995) was adapted to manufacturing problems.

The Product Model is separated into two parts: *graphic* and *structure of data*. The graphic part represents the structural decomposition of the product by using the concepts of *component*, *link* and *relation* proposed by our *Integrated Design Team* of the "Soils, Solids, Structures" Laboratory and recalled in (Tichkiewitch and Brissaud 2000) (Figure 2.18).

The second part regarding the structure of data gives the types of operations to be undergone by the product as well as the manufacturing parameters as viewed by the manufacturer.

To illustrate this, Figure 2.19 shows the graphic part of the guidance system of a vehicle.

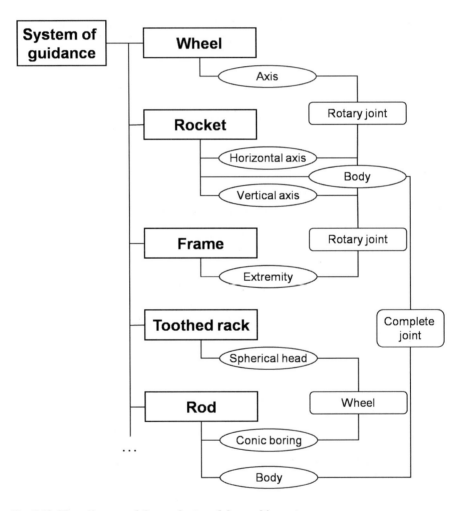

Fig. 2.19: Flow diagram of the product model: graphic part.

The next figure (Figure 2.20) shows a part of the operations which the product has to undergo at the various stations of machining such as turning, milling, drilling etc, as well as the parameters of these operations.

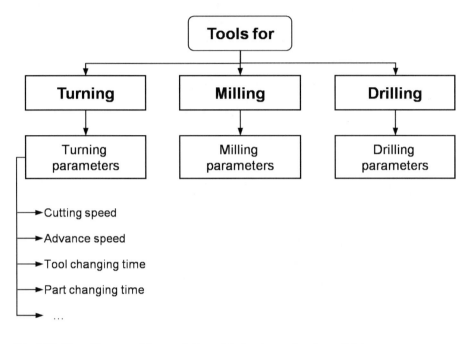

Fig. 2.20: Flow diagram of the product model structure: structure of data.

The product model is not static and can evolve in manufacturing in response to unforeseen elements (Figure 2.21, Figure 2.22). The parameters of design authorise additions and changes to the manufacturing parameters. The actors of manufacture are able to change the product model in agreement with the others actors of the design. The product model is unique in the fact that all modifications and additions can be read by all the actors concerned.

Consequently, it can be stated that the relation "order givers/subcontractor" changes. The credibility of a subcontractor does not only depend upon the quality of the products provided and the strict respect of deadlines but also upon his ability to react to inevitable dysfunctions. The industrial product of today "is more and more designed" by the subcontractor (supplier) and "less" by the client. Thus there is a need for:

- Reliable communication support to reactivate exchange information essential to management production,
- A model of the dysfunctions and risks in order to pre-empt their propagated effects.

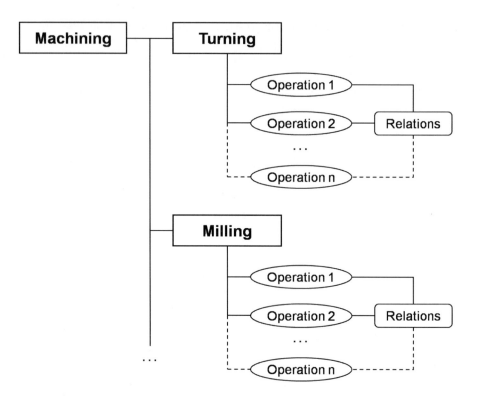

Fig. 2.21: Operation flow diagram of the manufacturing process.

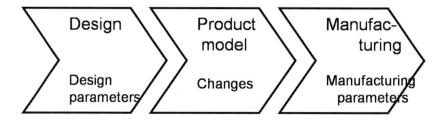

Fig. 2.22: Evolution of the product model.

Therefore, the production system becomes *reactive* and is characterised by a *synchronous approach* rather than interactive and characterised by an asynchronous approach. The impacts of these approaches at the organisational, economic and operational level of the installation also require assessment.

2.4.6 Product model from the market life cycle perspective

As product also consists of accompanying services, the model needs to be extended to become a core section with a surrounding shell. The proposed model for extended products takes the product perspective on the market and its evolution in the market context into consideration. Figure 2.23 shows the development of the product concept from the narrow sense to the broader one.

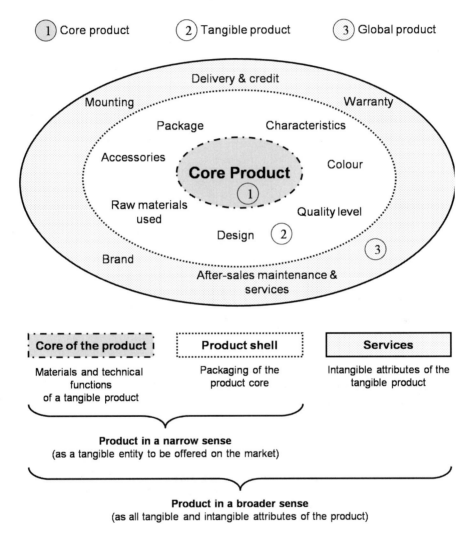

Fig. 2.23: Extended product model for the market life cycle.

The product model is based on the product life cycle and is used for elaborating a methodology for customer-orientated design.

2.4.7 Case study of a life cycle product model

This case study shows a product model from a life cycle perspective. The product model consists of the following types of attributes in order to incorporate variations in the service options during and at the end of the life cycle.

- Modularity: a product is modelled as a connectivity graph of modules. In turn, the modules are modelled as a connectivity graph of components forming the basic elements in this model (Figure 2.24). Modularity can be also used to explicitly classify the separation between the core and shell of a product.
- Life cycle related attributes: a component is modelled as a set of attributes, such as the options of recycling and reuse, cost of manufacture, manufacturing energy, lifetime, weight and material. The attributes define services which are available during and at the end of life cycles.

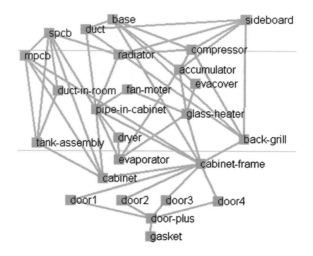

Fig. 2.24: Product model of a refrigerator (Umeda et al. 2000).

The model assumes that a module can be repaired by replacing defective components with working ones and that a component cannot be disassembled or repaired. In other words, used components which have been collected can be reused as they are, recycled or simply disposed of.

This modelling technique enables an appropriate modular structure to be designed based on the design parameter of products, modules and components such as usage, lifetime or material selection for recyclability. Figure 2.25 shows the optimised product models with respect to alternative life cycle types (reuse, maintenance and PMPP types) (Post Modern Manufacturing Paradigm). (Tomiyama, 1997) (Umeda et al. 2000)

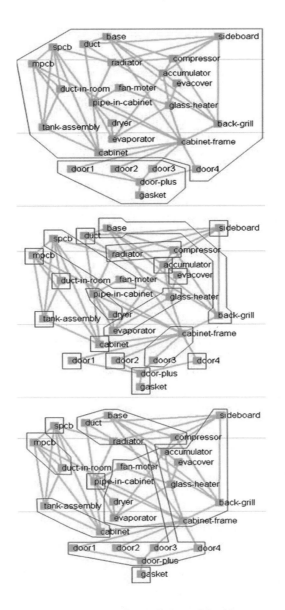

Fig. 2.25: Optimised module structures of a product model with respect to alternative live cycle types: Reuse type (above), maintenance type (centre) and PMPP type (below). (Tomiyama 1997), (Umeda et al. 2000).

2.5 Interim summary of life cycle modelling

Life cycle modelling represents the basis of a successful life cycle design. A number of methods, modelling techniques and international standards relevant to this can be found in literature and in industrial practice. The identification of factors affecting the life cycle and their adequately precise mathematical formulation is especially important in order to obtain valid information in a superordinate, holistic life cycle model.. Due to the long-term planning horizon, efficient models are required which depict the total life cycle of a product. The models need to be designed in such a way so that they are capable of showing various perspectives relevant to the situation concerned, thus permitting alternative product configurations and their development possibilities to be evaluated. Only then is it possible to identify holistic optimisation potentials in the early stages of product development and implement them to create a successful life cycle strategy.

2.6 References concerning chapter 2

(Abramovici 1997) Abramovici, M., Gerhard, D., and Langenberg, L., Application of PDM technology for product life cycle management, Proceedings of the 4th CIRP International Seminar on Life Cycle Engineering, Berlin, Germany, pp. 17-31, 1997.

(Al-Timini and MacKrell 1996) Al-Timimi, K., MacKrell, J., 1996, STEP: Towards Open Systems, CIMdata Inc., ISBN 1-889760-00-5. 2002.

(Aldanondo et al. 2000) Aldanondo, M., Reouge, S., Veron, M.: Expert configurator for concurrent engineering, Journal of Intelligent Manufacturing, 11(2), pp. 127–134, 2000.

(Andreasen 1992) Andreasen, M. M.: Designing on a "Designers Workbench" (DWB), 9th WDK Workshop, Rigi, Switzerland, 1992,

(Arkin 2002) Arkin, A., Business process modelling language, Technical report BPMI.org, 13 November, 2002

(Barkan 1988) Barkan, P.: "Simultaneous engineering", Design News, 44, A30, March 1988.

(Baumann 1990) Bauman, R.: CALS and Concurrent Engineering: business strategy or tool survival?, Aviation Week and Space Technology, 133, s1–s12, July 1990.

(Bhandarkar and Ngai 2000) Bhandarkar, M.P., Ngai, R.: "STEP-based feature extraction from STEP geometry for agile manufacturing", Computers in Industry, 41(1), pp. 3–24, 2000.

(Biren 1996) Biren, P.: Concurrent Engineering Fundamentals: Integrated Product and Process Organization, Prentice-Hall, Englewood Cliffs, NJ, 1996.

(Brissaud and Tichkiewitch 2001) Brissaud D., Tichkiewitch S., « Product Models for Life-Cycle », Annals of the CIRP, Manufacturing Technology, vol. 50, n° 1, 2001, p. 105-108.

(Buchholz and Franke 2003) Buchholz, A.; Franke, C.: Assessment of Standard Components for Extended Utilization. In: Proceedings Colloquium e-ecological Manufacturing, pp. 39–42, Berlin, 2003.

(Chan and Gu 1993) Chan, K.., Gu, P.: "A STEP based generic product model for concurrent engineering", Concurrent Engineering Methodology and Applications, pp. 249–275, 1993.

(Clark ad Fujimoto 1991) Clark, K.B., Fujimoto, T.: Product Development Performance, Harvard Business School Press, Boston, 1991.

(Collaborative Visions 2002) Collaborative Visions Inc., PLM user strategy, Technical report, (Rasmus 1993)   Rasmus, D.: Learning the waltz of synthesis, Manufacturing Systems, 11(6), pp. 16–23, 1993.

(Fangyi et al. 2002) Fangyi, L., Guanghong, D., Jinsong, W., Dong, X., Xueping, L., Peng, M., Wei, M., and Sa, L., Green design assessing system model of products, Proceedings of the 2002 IEEE International Symposium on Electronics and the Environment, pp. 123-127, 2002.

(Fenves et al. 2003) Fenves, S. J., Sriram, R. D., Sudarsan, R., and Wang, F., A product information modeling framework for product lifecycle management, Proceedings of the International Symposium on Product Lifecycle Management, Bangalore, India, 16-18 July, 2003.

(Goncharenko et al. 1999) Goncharenko, I., Kryssanov, V. V., and Tamaki, H., An agent-based approach for collecting and utilizing design information throughout the product life cycle, Proceedings of the 7th IEEE International Conference on Emerging Technologies and Factory Automation (ETFA1999), 1, pp. 175-182, 18-21 October, 1999.

(Gruzien 2002) Grudzien, W.: Beitrag zur Steigerung der Nutzenproduktivität von Ressourcen durch eine Life Cycle Unit. Dissertation TU Berlin, 2002.

(Harrington 1973) Harrington, J.: Computer Integrated Manufacturing, Kal Krieger Publishing, Malabar, 1973.

(Hong Bae et al. 2005 a) Hong Bae Jun, Dimitris Kiritsis, Xirouchakis, P. Product lifecycle modeling with RDF, Proceedings of International Conference on Product Lifecycle Management (ICPLM' 05), IUT Lumiere-Lumiere University of Lyon, France, 11-13 July 2005.

(Huang and Mak 1999) Huang, G.Q., Mak, K.L.: "Web-based morphological charts for concept design in collaborative product development", Journal of Intelligent Manufacturing, 10(3), pp. 267–278, 1999.

(IS0 1998) ISO/TC184/SC4, 1998, Guidelines for the development and approval of STEP application protocols, International Organization for Standardization, SC4N535.

(ISO 1999) ISO/TC184/SC4, 1999, Industrial Automation Systems and Integration – Product Data Representation and Exchange – Part 235: Application Protocol: Materials Information for Product Design and Validation, International Organization for Standardization, ISO/NWI 10303-235.

(ISO 2000) ISO/TC184/SC4, 2000, Industrial Automation Systems and Integration – Product Data Representation and Exchange – Part 236: Application Protocol: Furniture Product Data and Project Data, International Organization for Standardization, ISO/NWI 10303-236.

(ISO 2005a) ISO/TC184/SC4, 2005, Industrial Automation Systems and Integration – Product Data Representation and Exchange – Part 239: Application Protocol: Product Life Cycle Support, International Organization for Standardization, ISO/CD 10303-239.

(ISO 2005b) ISO/TC29/WG34, 2005, Cutting Tool Data Representation and Exchange – Part 1: Overview, Fundamental Principles and General Information Model, International Organization for Standardization, ISO/IS 13399-1

(Iwata et al. 1995) Iwata, K., Onosato, M., Teramoto, K., Osaki, S.A.: Modeling and Simulation Architecture for Virtual Manufacturing Systems, Annals of CIRP, 44(1), pp. 399–402, 1995.

(Johansson 2001) Johansson, M.: Information Management for Manufacturing System Development, Doctor Thesis, Computer Systems for Design and Manufacturing, KTH, 2001.

(Kimura and Suzuki 1995) Kimura, F. and Suzuki, H., Product Life Cycle Modelling for Inverse Manufacturing, Proceedings of the IFIP WG5.3 International Conference on

Life-cycle Modelling for Innovative Products and Processes, Berlin, Germany, pp. 80-89, 1995.

(Kind 2000) Kind, Chr.: Demontageorientierte informationstechnische Infrastruktur. Tagungsband „Kolloquium zur Kreislaufwirtschaft und Demontage" des Sonderforschungsbereichs 281 „Demontagefabriken zur Rückgewinnung von Ressourcen in Produkt- und Materialkreisläufen", Berlin, 20./21. Januar 2000, S. 84–87.

(Kiritsis 2004) Kiritsis, D., Ubiquitous product lifecycle management using product embedded information devices, Proceedings of International Conference on Intelligent Maintenance Systems (IMS 2004), 2004.

(Kiritsis and Rolstadås 2005) Kiritsis, D. ,Rolstadås, A., PROMISE-A closed-loop product lifecycle management approach, Proceedings of IFIP 5.7 Advances in Production Management Systems: Modeling and implementing the integrated enterprise, 2005.

(Kiritsis et al. 2003) Kiritsis, D., Bufardi, A., Xirouchakis, P.: Research issues on product lifecycle management and information tracking using smart embedded systems, Advanced Engineering Informatics 17(2003) 189-202.

(Klausner and Grimm 1998) Klausner, M. and Grimm, W. M., Sensor-based data recording of use conditions for product takeback, Proceedings of the 1998 IEEE International Symposium on Electronics and the Environment, pp. 138-143, 1998.

(Krause and Schlingheider 1993)   Krause, F.L., Schlingheider, J.: Product modeling, Annals of CIRP, 42, pp. 695–706, 1993.

(Krause 2004) Krause, F.-L., Kind, Chr., Jungk, H.: Product Condition Model. In: "Design In The Global Village" 14th International CIRP Seminar 2004, Cairo, Egypt 16./18. Mai 2004

(Lubell et al. 2004) Lubell, J., Peak, R. S., Srinivasan, V., and Waterbury, S. C., STEP, XML, and UML: Complementary technologies, Proceedings of the DETC 2004: ASME 2004 Design Engineering Technical Conferences and Computers and Information in Engineering Conference, Utah, USA, 2004.

(Matheus 1995) Matheus, J.: Organizational foundations of intelligent manufacturing systems – the holonic viewpoint, Computer Integrated Manufacturing Systems, 8(4), pp. 237–243, 1995.

(Martini 2000) Martini, K.: Simulationswerkzeuge zur demontagegerechten Produktentwicklung. Tagungsband „Kolloquium zur Kreislaufwirtschaft und Demontage" des Sonderforschungsbereichs 281 „Demontagefabriken zur Rückgewinnung von Ressourcen in Produkt- und Materialkreisläufen", Berlin, 20./21. Januar 2000, S. 80–83.

(Ming and Lu 2003) Ming, X. G. and Lu, W. F., A Framework of Implementation of Collaborative Product Service in Virtual Enterprise, Innovation in Manufacturing Systems and Technology (IMST), http://hdl.handle.net/1721.1/3740, January, 2003.

(Mitrouchev et al. 1998) Mitrouchev P., Brun-Picard D., Hollard M., Haurat A., « A New Product-Model for Production », *Proceeding of 2-nd International Conference on Integrated Design and Manufacturing in Mechanical Engineering, I.D.M.M.E.'98*, May 27-29, 1998, Compiègne, ISBN: 2-913087-03-5, vol. 4, p. 1179-1186.

(Mortensen and Andreasen 1996) Mortensen, N.H., Andreasen, M.M., Designing in an Interplay with a Product Model, explained by design units, TMCE'96 Budapest Hungary, 1996

(Murphy 1950) Murphy, G.: Similitude in Engineering, The Ronald Press Company, New York, 1950.

(Nagal and Dove 1991) Nagal, R., Dove, R.: 21st Century Manufacturing Enterprise Strategy: An Industry-Led View and Infrastructure, Iacocca Institute, Lehigh University, 1991.

(Nielsen 2003) Nielsen, J., 2003, Information Modeling of Manufacturing Processes: Information Requirements for Process Planning in a Concurrent Engineering Environment, Ph.D. thesis, Kungliga Tekniska Högskolan, ISSN 1650-1888.

(Niemann 2003a) Niemann, J. (2003): Ökonomische Bewertung von Produktlebensläufen-Life Cycle Controlling. In: Bullinger, Hans-Jörg (Hrsg.) u.a.: Neue Organisationsformen im Unternehmen : Ein Handbuch für das moderne Management. Berlin u.a. : Springer, p. 904-916

(Niemann 2003b) Niemann, J.: Life Cycle Management, In: Neue Organisationsformen im Unternehmen - Ein Handbuch für das moderne Management, Bullinger, H.-J., Warnecke, H. J., Westkämper E. (Ed.), 2. Auflage, Springer Verlag, Berlin u. a.; 2003

(Niemann 2007) Niemann, J. 2007. Eine Methodik zum dynamischen Life Cycle Controlling von *Produktionssystemen*. Stuttgart, Germany: University of Stuttgart (Dissertation). Heimsheim, Germany: Jost-Jetter.

(Niemann et al. 2004) Niemann, Jörg; Stierle, Thomas; Westkämper, Engelbert: Kooperative Fertigungsstrukturen im Umfeld des Werkzeugmaschinenbaus : Ergebnisse einer empirischen Studie. In: Wt Werkstattstechnik 94 (2004), Nr. 10, S. 537-543

(Niemann and Westkämper 2005) Niemann, J., Westkämper, E. (2005) : Dynamic Life Cycle Control of Integrated Manufacturing Systems using Planning Processes Based on Experience Curves. In: Weingärtner, Lindolfo (Chairman) u.a., CIRP: 38th International Seminar on Manufacturing Systems / CD-ROM: Proceedings, May 16/18 - 2005, Florianopolis, Brazil. p. 4

(Nonomura et al. 1999) Nonomura, A., Tomiyama, T., and Umeda, Y., Life cycle simulation for inverse manufacturing, Proceedings of the 6th international seminar on life cycle engineering, pp. 304-313, 1999.

(Onosato and Iwata 1993) Onosato,M., Iwata, K.: Development of a Virtual Manufacturing System by Integrating Product Models and Factory Models, Annals of CIRP, 42(1), pp. 475–479, 1993.

(Ou-Yang and Pei 1999) Ou-Yang, C., Pei, H.N.: "Developing a STEP-based integration environment to evaluate the impact of an engineering change on MRP", International Journal of Advanced Manufacturing Technology, 15, pp. 769–779, 1999.

(PROMISE 2004) PROMISE, PROMISE-Integrated project: Annex I-Description of Work, Project proposal, 2004.

(PROMISE 2005) PROMISE, DR2.1 PROMISE generic model (version 1), Technical report, 2005. Proceedings of the IFIP WG5.3 International Conference on Life-cycle Modelling for Innovative Products and Processes, pp. 43-55, 1995.

(Qaissi et al. 1999) Qaissi, J.H., Coulibaly, A., Mutel, B.: Product data model for production management and logistics, Computers and Industrial Engineering, 37(1–2), pp. 27–30, Oct. 1999.

(Scheer 1998a) Scheer, A.-W., ARIS Business process framework, Springer, 1998a.

(Scheer 1998b) Scheer, A.-W., ARIS-Business process modeling, Springer, 1998b.

(Scheidt and Zhong 1994) Scheidt, L. and Zong, S., An approach to achieve reusability of electronic modules, Proceedings of 1994 IEEE International Symposium on Electronics and the Environment (ISEE 1994), pp. 331-336, San Francisco, USA, 2-4 May, 1994.

(Schneider and Marquardt 2002) Schneider, R., and Marquardt, W., Information technology support in the chemical process design life cycle, Chemical Engineering Science, 57(10), pp. 1763-1792, 2002.

(Seltes 1978) Seltes, J.W.: "A feature-based representation of parts for CAD", BS Thesis, Mechanical Engineering Department, MIT, 1978.

(Terzi et al. 2004) Terzi, S., Panetto, H., and Morel, G., Interoperability standards along the product lifecycle: a survey, Proceedings of International IMS forum 2004: Global challenges in Manufacturing, pp. 925-934, Italy, 2004.

(Tichkiewitch et al. 1995) Tichkiewitch S., Chapa E., Belloy P., « Un modèle produit multi-vues pour la conception intégrée », *Congrès international de Génie Industriel de Montréal, Montréal (Canada),* Octobre 1995, p. 95-129.

(Tichkiewitch and Brissaud 2000) Tichkiewitch S., Brissaud D., « Co-ordination Between Product and Process Definitions in a Concurrent Engineering Environment », *Annals of the CIRP, Manufacturing Technology*, vol. 49, n° 1, 2000, p. 75-78.

(Tipnis 1995) Tipnis, V. A., Toward a comprehensive life cycle modeling for innovative strategy, systems, processes and products/services, (Kimura 2002)   Kimura, F. A computer-supported approach to life cycle design of eco-product, Technical report, 2002.

(Tomiyama 1997) Tomiyama, T.: A manufacturing paradigm toward the 21st century. Integrated Comput. Aided Eng. 4, pp. 159–178., 1997

(Trappey et al. 1997) Trappey, J.C., Liu, T.H., Hwang, C.T.: Using EXPRESS data modeling technique for PCB assembly analysis, Computers In Industry, 34(1), pp. 111–123, 1997.

(UGS PLM 2004) UGS PLM co., Open product lifecycle data sharing using XML, PLM XML white paper, http://www.ugs.com/products/open/plmxml/downloads.shtml, 2004.

(Umeda et al. 2000) Umeda, Y., Nonomura, A. and Tomiyama, T.: A Study on Life-Cycle Design for the Post Mass Production Paradigm, Artificial Intelligence for Engineering Design, Analysis and Manufacturing, Vol.14, No. 2, Cambridge University Press, pp. 149-161., 2000

(WfMC 2002) WfMC, Workflow management coalition standard: workflow process definition interface - XML process definition language, Version 1.0, Technical report WfMC TC-1025, WfMC, 25 Oct 2002

(Winkler and Mey 1994) Winkler, J., Mey, M.: Holonic manufacturing system, European Production Engineering, 19(3), pp. 10–12, 1994.

(Winner 1988) Winner, R.I.: "The role of concurrent engineering in weapons system acquisition", IDA Report R-338, Institute of Defense Analysis, Alexandria, VA, 1988.

Womack et al. 1989) Womack, J.P., Jones, D.T., Roos, D.: The Machine that Changed the World, Macmillan, 1989.

(Youssef 1992) Youssef, M.A.:Agile manufacturing: a necessary condition for competing in globe markets, Industrial Engineering, pp. 18–20, 1992.

(Zhao 1998) Zhao, Y.S: "The STEP based multi view integrated product modelling", The 9th Symposium on Information Control in Manufacturing, France, pp. 345–349, 1998.

3 Life cycle evaluation

The previous chapter presented the various approaches in the field of life cycle modelling. This chapter is concerned with the economical and ecological acquisition and assessment of product life cycles (see Figure 3.1). As life cycle costs are especially relevant in industrial practice, particular attention has been paid to this aspect. By recording, analysing and optimising these costs, significant potential economic benefits can be tapped during the product life cycle.

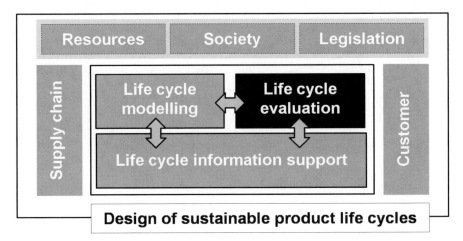

Fig. 3.1: Structure of chapter 3.

It is not just due to rising energy prices, the tightening of statutory regulations and increasing consumer awareness that products in the future will be more and more carefully examined and assessed with regard to their potentials for improvement and ecological impact. It is essential that the design sustainable of product life cycles considers and optimises both ecological and economical effects. As a result, the following chapter presents standards and innovative concepts from research and practice.

3.1 Economical assessment of product life cycles

The following graph (Figure 3.2) represents the principal course of a product's value over its lifetime. During the comparatively short production phase, product value rises to sale price. Subsequent to wear and increasing failure rates, a drop in value occurs which can be partially compensated through maintenance, repair and upgrading. Once no further measures are possible, recycling at least enables the material value to be retained. Higher values can only be achieved through reuse and re-manufacturing.

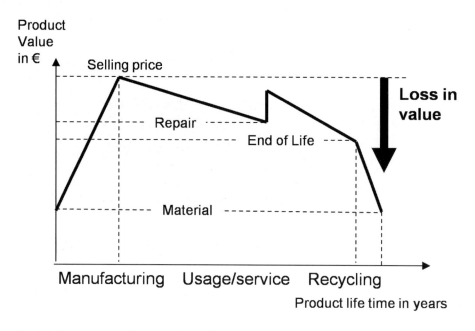

Fig. 3.2: Product value during its life cycle.

The deterioration of product value does necessarily mean that the overall value of all modules and components belonging to the product is lost. The utilisation of reusable components offers high utilisation efficiency as a whole. As a result, a *new life* with conservation of the material and (at least a part of) the economic value of a product can begin once again.

The economical feasibility of measures to maintain value and to recuperate remaining values at the end of a product's life is a function of product value at that particular point in the product's life cycle. Life cycle cost (LCC) methods can be used to calculate the product value and to pro-actively assess future developments. LCC can be defined as a systematic analytical process for evaluating various designs or alternative courses of action with the objective of choosing the best way to employ scarce resources. (Kumaran et al. 2003)

Pro-active assessment of future cost developments requires the in-depth knowledge of possible developments. Costs (Blanchard 1978) measured over an entire life cycle possess the structure of an iceberg (Figure 3.3).

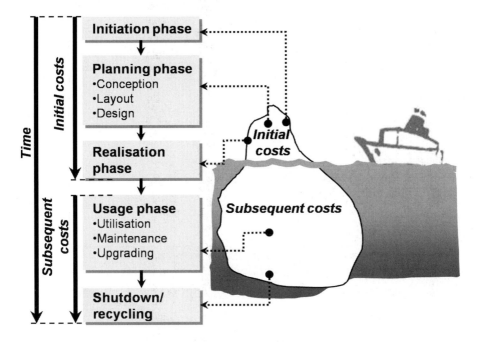

Fig. 3.3: The iceberg effect (Blanchard 1978), (Wübbenhorst 1992).

Arising costs can be divided into initial costs and subsequent costs over the lifetime. As with an iceberg, only a small part of the entire (cost) block is visible to start with. The main part is (initially) hidden but must be taken into account in order to avoid "shipwreck". By assigning the various costs to specific life cycle phases, it becomes clear that the initial costs occur during the phases of initiation, planning and realisation. The product is manufactured in this period. Subsequent costs mainly arise in the usage phase of a product. Due to the fact that most of the financial transactions are performed during this phase, an estimation of costs and revenues is essential. Studies have shown that a focus on minimised initial costs by ignoring (later) subsequent costs does not result in minimised total life cycle costs (Wübbenhorst 1992). Nevertheless, subsequent costs are essentially determined in the early phases of product design. One way of evaluating and analysing these economic correlations is to apply the method of life cycle costing.

3.1.1 Life cycle costing

The main objective of a life cycle cost analysis (LCC analysis) is to maximise the difference between life cycle costs and benefits (finally evaluated as profits).

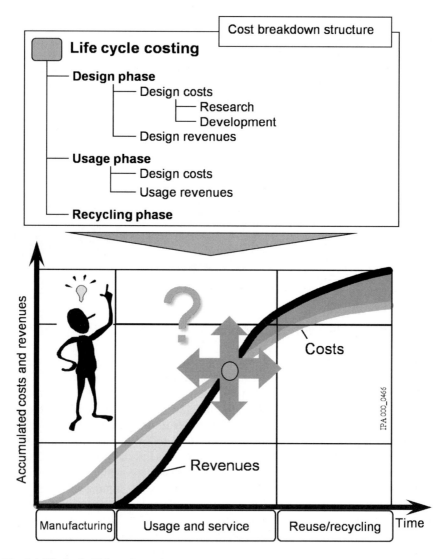

Fig. 3.4: Method of life cycle costing.

In the process, life cycle costs can be roughly divided into the three segments of development costs, utility /service costs and recycling/ reprocessing costs. Figure 3.4 shows an example of such a cost breakdown structure.

The passing through of these three phases is as important as the technical product life cycle itself. Analogous to costs, life cycle revenues can also be allocated to the individual phases in a similar way. Firstly, relevant cost and revenues blocks are recorded; the type of the individual positions varies depending on the investment goods investigated. To assess total success in the life cycle, the positions are aggregated separately according to expenditure and revenue spread out over all the phases. If only marginally-different functional solution options exist for a certain system design, a life cycle cost analysis may be of assistance in finding the best economical variant. If – in later phases – the results of the life cycle cost analysis carried out are critically assessed from time to time and are used to optimise the design within an improvement cycle, a technically and economically optimised product is the final result. The method can be applied both to plan and identify profit potentials over the entire product life cycle:

- Calculation of total costs for products (initial and subsequent costs)
- Identification of cost and revenue drivers
- Impact of outsourcing decisions
- Cash-flow analyses, return on investment (ROI)
- Analysis of "what if" scenarios (prognosis)
- Optimal due-date for machine replacement
- Holistic investment budgeting
- Analysis of trade-offs between initial and subsequent costs
- Analysis of customer lifetime value (evaluating the value/profitability of each business relation)

By looking at the situation in the long-term, LCC analyses uncover hidden cost drivers as well as profit potentials during the entire life cycle. In this way, the analysis also supplies parameters for outsourcing strategies right up to calculations for modern full-service concepts and complete outsourcing. LCC is also consistent with "Design for X" approaches, where the perception that decisions made in the earliest design step have important and irreversible effects upon the whole life cycle of the designed product.

For these reasons, LCC helps companies to pursue PLM objectives and can be considered as part of the initial phases making up a PLM approach. LCC can form part of the portfolio evaluation of new concepts and new ideas before starting the functional design (see also chapter 6.7). Adopting a LCC approach for assessing and evaluating innovations also requires the adoption of a set of procedures and tools which allow data to be systematically collected and stored.

Exact estimation of such financial figures accumulated during the entire product life cycle inevitably requires information about the behaviour of products and the actors involved in the life cycle (i.e. users, producers and service providers). To obtain this information, life cycle simulation techniques based on digital data are necessary.

3.1.2 Strategic portfolio for optimising life cycle costs

On analysing the life cycles of various products, it becomes obvious that the life span is of different duration. The products also differ in their degree of complexity and product value. A strategy to master and optimise product life cycles has to take this into account. This implies that no one universally-valid optimisation strategy exists. Also, a differentiated approach is required in order to identify the key cost drivers. To achieve this, the cost breakdown structure of a product life cycle can be divided into the general categories of investment and running costs. The ratio of these two cost drivers helps to identify the variables for optimisation.

The investment costs cover all costs which occur only once, e.g. cost of design, installation, training and recycling. Running costs comprise all the operational costs (including maintenance, service, unplanned downtimes, etc.) incurred during the usage of a product throughout its entire life cycle. It is essential that both types of costs are recorded (or at least forecast) over the total life cycle. The cause of specific cost origins may be anchored in various points within the life cycle and it is often a highly time-consuming and problematic process to identify them. A detailed analysis of trade-offs may be of assistance here (Figure 3.5).

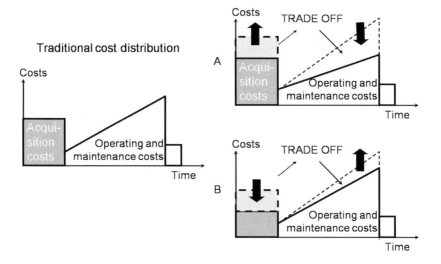

Fig. 3.5: Strategic courses of action to optimise a life cycle. (extending according to Wübbenhorst, K. 1984)

On considering the trade-offs and methods of coordinating the entire life cycle, it can be seen that the phases are not independent of one another. The product realisation phase plays a particularly important role in life cycle costing. As ascertained previously, this is because a large part of the costs are fixed in the early phases of a product's life cycle which cannot be influenced later on in the cycle.

Therefore, decisions need to be made at an early stage which take interrelationships between the various quality, time, profit and cost aims into account. The impact of these decisions on product success can be accurately analysed using the instrument of life cycle costing.

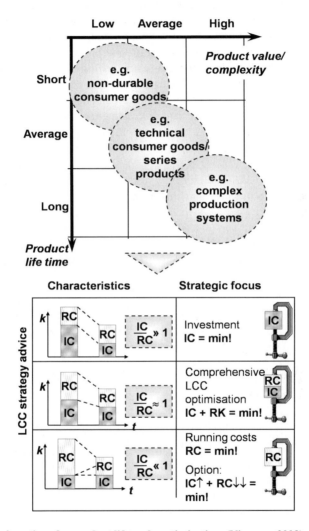

Fig. 3.6: Strategic options for product life cycle optimization. (Niemann 2003)

In order to make informed decisions, integrating and long-term viewpoints are required and the individual aspects of each aim have to be made quantifiable. In a second step, the strategy advice is assigned to specific product functions to enable concrete actions to be executed at the base of cost origins.

As Figure 3.6 shows, three categories with varying time scales and strategies may thus be defined. The first group is characterised by a short lifetime and low product value or complexity. Such non-durable technical consumer goods are usually mass-produced and manufactured in large series. The main emphasis of life cycle management is placed on the rational organisation of services, marketing and product recycling techniques. Robust techniques can be used for recycling due to the fact that the added value profit is low in relation to the value of the product. The second category is assigned to series products with a limited number of variants. Life cycle management of these products includes services and maintenance as well as industrial recycling and the partial reuse of parts and components. High-quality capital goods are assigned to the third category. The main emphases here are on maximum utilisation strategies, the maintenance of performance and additional added value in the field of after-sales. Industrial recycling only plays a minor economic role in this category of products. (Brussel and Valckenaers 1999)

3.1.3 Standardised worksheet for evaluating life cycle costs

More and more customers are asking their manufacturers about the total cost of ownership for investment goods. As a result, the total costs become a focus and have to be determined and made transparent by the manufacturer at the point of sale.

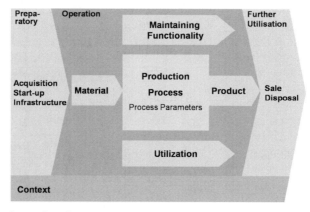

Fig. 3.7: Life cycle cost functions.

To do this, a standardised cost sheet is necessary which covers all relevant cost positions over the life cycle. The general framework for this is shown in Figure 3.7. The cost sheet is structured according to system theory and uses a functional approach to generate a consistent cost breakdown structure over the product life cycle. The Federal Association of Engineering (VDMA, Germany) started up a working group to create an industrial standard in this field. With its extensive knowledge and experience, the working group within KCiP has contributed to the development of such a standardised sheet. A catalogue of over 60 parameters covering all life cycle phases has been compiled and integrated into the scheme and includes all the necessary arithmetic operations. A screenshot of this structure is shown in Figure 3.8.

Table 3 — Cost Elements of the Preparatory Phase

Code	Name	Description	Calculation Formula	Unit
E	Preparatory costs	Acquisition and infrastructure costs.	E1 + E2 + E3	Currency
E1	Acquisition costs	All costs incurred during the preparatory phase prior to start of production.	E1.1 + E1.2 + E1.3 + E1.4 + E1.5 + E1.6 + E1.7 + E1.8 + E1.9 + E1.10	Currency
E1.1	Acquisition price	Price of the machine, including legal warranty.	Input	Currency
E1.2	Initial set of tools	Price of the tools that are purchased along with the machine.	Input	Currency
E1.3	Service parts package	Price of the service parts that are purchased along with the machine.	Input	Currency
E1.4	Warranty extension	Price for warranty extension in accordance with requirements. Negative value in the case of warranty reduction.	Input	Currency
E1.5	Installation costs	Total.	E1.5.1+E1.5.2+E1.5.3+E1.5.4	Currency
E1.5.1	Personnel costs: installation	Expenses for personnel required for installation	Input	Currency

Fig. 3.8: Screenshot of the cost breakdown structure (VDMA 2006).

The scheme will now serve as a standardised basis for the calculation of life cycle costs among European manufactures (see also Chapter 6.7.1).

3.1.4 Case study of a life cycle cost calculation for a machine tool

The following paragraphs describe the case of a machine tool builder - here a transfer machine tool constructor - who is requested by an important client to calculate and include the life cost of a particular machine in the offer. Transfer machines are utilised to manufacture a particular part (or a particular family of parts). The manufacturer has to demonstrate that his machine can produce a certain part at a requested production rate and quality level demanded by the buyer. These quality and production levels have to be maintained during the entire life of the machine tool which is estimated to be 10 years. LCC calculations are usually based on this figure for the total duration of machine life.

These machines need to be available as much as possible during their lifetime. Breakdowns are directly related to loss of production. The client belongs to the automotive sector and is particularly worried about the issue of machine availability. The client is well aware of the importance of knowing the life cycle cost and does not want to acquire machine tools, or equipment in general, based only on production capacity and price. Customers want a guarantee of machine availability. Even beyond availability, they want a machine which is relatively cheap to use in order to compensate for high acquisition costs. This is also a crucial issue for European machine tool builders if they want to keep their position in the market.

If they succeed in producing reliable machines which consume less, it would be possible to convince purchasers to make a higher investment when buying machines. To achieve this, manufacturers have to be able to prove to the buyer that the life cycle cost of their machines is lower and has been calculated based on objective data, i.e. real machine performance.

Buyers may soon start requesting contractual commitment to life cycle cost. For certain products such as trains, for example, it is not unusual for the purchaser to buy products and services such as maintenance at the same time. Therefore, it is essential the manufacturer is precisely informed about the performance of his products throughout their entire life cycle. The product life cycle model used by the manufacturer is one as simple as the one shown in Figure 3.9 below:

Fig. 3.9: Life cycle cost model.

The cost corresponding to the construction phase is called the *acquisition cost* and comprises the following parts:
- Purchase price
- Administration costs
- Installation
- Training
- Shipping cost
- Warranty
- Support equipment

This is what the buyer has to spend when he acquires and installs the machine tool. The manufacturer may provide some of these services (such as training). It is a matter of negotiation between the manufacturer and customer as to who will bear the remaining costs (e.g. shipping costs). In any case, these are costs which do not require any special calculations but are fixed at the time of purchase of the machine tool. Operating costs are those incurred when using the machine in normal working conditions and include the following:
- Tooling
- Direct labour
- Changeover labour
- Consumables
- Utilities
- Waste handling
- Floor space

Tooling costs are calculated based on tool wear. As the operations to be carried out are well known, the manufacturer has to provide the number of parts which can be produced by each tool. As the production rate is known, it is possible to deduce the number of tools which will be required during the life of the machine tool (usually set at 10 years). The cost would therefore be the cost of each individual tool times the number of tools used. Using this method, the total number of tools is calculated assuming the machine will be working the total number of hours allocated to it., However, as machines do not have 100 % availability in reality, the operational availability of the machine tool therefore has to be calculated.

Availability is the ratio between the uptime and the total allocated working time for the machine. Downtime is the addition of the times when the machine is stopped due either to a machine breakdown or because the machine user is simply not using it for organisational reasons (e.g. lack of input parts). Inherent availability is then obtained from the ratio between machine downtime and the total time. Finally, operational downtime is the ratio between the total downtime and the total time. The manufacturer should be able to estimate the inherent availability of his machines from the historical data of breakdowns of his machines. Thus, the actual number of working hours is obtained by multiplying the operational availability by the total number of hours.

Similarly, in order to calculate labour costs, information from both the user (hourly labour cost, number of working hours/year) and the manufacturer (number of operators per machine) is required. Consumables and utilities refer to costs of materials and fluids required by the machine to function normally. They are grouped as follows. Consumables:

- Coolant
- Filters
- Lubrication
- Utilities:
- Compressed air
- Electricity
- Gas
- Steam
- Water

A consumption rate is proposed for all these cost concepts which is then multiplied by the number of hours which the machine is actually available. Costs also have to be calculated for waste handling and the amount of floor space occupied by the machine. Waste handling comprises:

- Coolant dumping
- Filter
- Sludge

Costs of coolant and sludge are given per litre. Costs of disposing of filters is expressed per metre. No other environmental impact costs are foreseen (except disposal costs, if they can be considered as such). The second large group of usage costs is those termed maintenance costs. Here, three types of costs are considered:

- Preventive maintenance costs
- Unscheduled maintenance costs
- Spare parts costs

PM costs are a function of planned preventive maintenance tasks and their duration. Corrective or unscheduled maintenance is a function of the number of stops due to breakdowns and the time required to repair. The manufacturer has to provide the MTBF (Mean Time Between Failures) and the MTTR (Mean Time To Repair) of the equipment. These need to be derived from historical data of equipment already installed which is equal or "similar" to that being analysed. These parameters can be calculated by adjusting failure and repair distributions using a Weibull function.

Spare part costs are also dependent on the failure rate, the latter also having to be derived from historical data. Finally, the cost of disposal or retrofitting has to be taken into account in order to complete the life cycle cost.

In transfer lines such as that mentioned in this case, it is quite usual that acquisition costs, operation costs and maintenance costs each represent 1/3 of the total costs (without the consideration of disposal costs).

3.1.5 Continuous life cycle cost controlling

Today, manufacturing companies in the field of high-variant series productions operate in markets which are extremely dynamic and turbulent. Here especially, changing customer demands create a challenge for companies to manufacture high-quality products and variants within shorter and shorter innovation cycles and to launch them onto the market. (Spath 2003), (Westkämper 2004), (Hummel and Westkämper 2006)

According to calculations made by the Federal Statistical Office, technical investment goods with a (new) value of almost 12 billion Euros were utilised for industrial manufacturing in Germany in 2007 (Federal Statistical Office 2007). In the field of series productions, these investments were mainly made in direct relation to the product being manufactured. This shows that the economic calculation of a capital investment needs to be orientated towards the planned turnover curve of manufactured products. The expected life cycle (market cycle) of products over time thus essentially determines the type and amount of investment made. However, as far as manufacturing companies are concerned, the presence of market change drivers means that the inflow of future orders is affected by increasing planning uncertainty with regard to the type (product mix) and number of products being manufactured.

In the process, the main objectives of a company with regard to efficiency are defined by a *magical triangle* formed by cost reduction, time reduction and quality improvement. Series manufacturers work under high cost pressures with the result that the efficiency of manufacturing systems is usually expressed in piece costs. The aims of reducing times and improving quality are forced to accept a subordinate role here and thus be converted to piece cost values. Consequently, companies are constantly being challenged to manufacture product units at the lowest possible piece costs in order to keep up with the competition. As far as the effective usage of operating manufacturing systems is concerned, this results in the problem of having to constantly absorb external adaptation pressures by making continuous internal adjustments. In this context, a manufacturing system is seen as being an independent unit in which all resources required for manufacturing a specific product are united. (Niemann 2007), (Westkämper 2003, 2006)

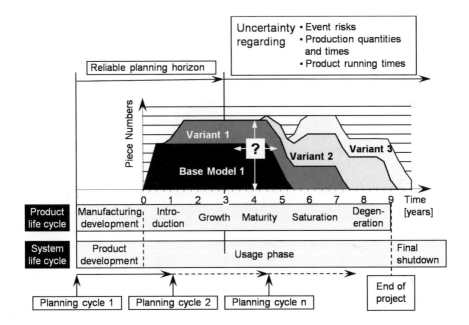

Fig. 3.10: Continuous adaptation of a manufacturing system and product spectrum. (Niemann 2007)

In general, when planning a product or manufacturing plant, precise planning data is only available for a limited period of time. Planning reliability concerning future order quantities and their composition is only present later on. For this reason, the planning measures taken only once at the start of an investment tend to be highly unreliable. Therefore, systems need to be reconfigured and optimised at regular intervals as more and more concrete information is obtained. Figure 3.10 illustrates this problem in the form of a graph. Technical advances also demand constant innovation and this is reflected in the development of new machines either with improved efficiency or with an increased scope of performance.

Consequently, technical, economical and organisational change drivers need to be constantly adapted - especially as far as high-capital investment goods are concerned - in order to achieve optimum machine usage throughout the life cycle of a system. (Niemann 2007), (Spath 2003)

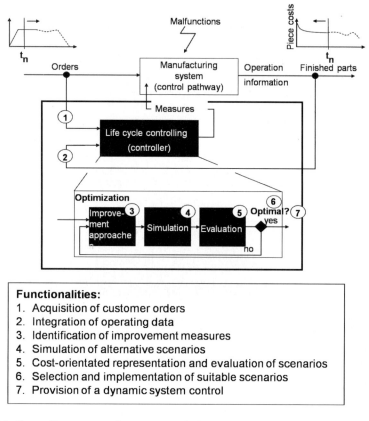

Fig. 3.11: Controlling concept for system optimisation based on a control cycle. (Niemann 2007)

There is therefore a requirement to develop a method which supports the continuous optimisation of manufacturing systems throughout their life cycle in the field of series productions. The method has to be capable of recording the effects of central change drivers on manufacturing systems and of providing systematic control mechanisms for an adapted system reconfiguration in order to ensure production with optimised piece costs in dependence upon the output quantities required. In this book, this control model has been called *life cycle controlling*. (Niemann 2007) Here, the main aspects of controlling include not only system-design functions but also system-monitoring functions, especially an anticipatory acting influence with regard to continuous system optimisation. Figure 3.11 shows this control concept with the necessary sub-functions as a graph.

3.1.5.1 Design concepts for manufacturing systems

The term *manufacturing* characterises the "entire economical, technological and organisational measures involved in the development and processing of materials" (Westkämper 2006). In the process, the term manufacturing in accordance with Westkämper embraces much more than just manufacturing, because it not only includes the "manufacturing processes themselves but also all controlling and organisational functions associated with them [...], from development right up to delivering a product to the customer" (Westkämper 2004). Based on this definition, the following existing design concepts for manufacturing systems can be fundamentally classified into: organisational concepts (e.g. Eversheim 1999); organic concepts (e.g. Warnecke 1993); system-related concepts (e.g. Westkämper 2004, 2006) and method-orientated concepts (e.g. Spath 2003).

The analysis of design concepts shows that today's manufacturing systems for series productions have been designed as closed autonomous units. System operation is supported by standardised process sequences and is subject to integrated optimisation attempts with regard to cost, time and quality. The system concept of the Stuttgart company model (Westkämper 2006) is especially suitable as a basis for modelling and representation. The model is characterised by a hierarchical system set-up and performance units similar to itself. Via this set-.up, dynamic alterations to a system can be depicted and described throughout its life cycle. By superimposing methodical concepts, rationalisation measures can be addressed and evaluated with regard to their cost effectiveness. In order to be able to continuously update the actual planning situation, feedback from real company processes is required. This has not been realised in the past in concepts known to date due to the complexity of interrelated effects. As a result, insufficient direct economical feedback has taken place during continuous system monitoring.

Deficits occur as a result of failing to integrate planning and operating data into a life cycle orientated planning system. Such data would enable adaptations and the necessary alterations to a manufacturing system to be pro-actively planned in order to increase its usage in dependence upon a volatile manufacturing programme and to permit the effects (or benefits) to be verified in advance. To achieve this, existing planning data from the rough planning phase from PPS or ERP systems could be utilised which would enable manufacturing orders to be acquired right up to the planning horizon. With current planning systems, order control takes place under the primacy of optimised resource allocation for a defined manufacturing system. However, as the design of a manufacturing system is the object of the control model to be created, a suitable method needs to be developed to resolve manufacturing orders. This would enable a system to be optimally configured based on this. In order to analyse life cycle related potentials (trade-offs), the progression of piece costs over time needs to be investigated over the later life cycle of the system. Due to the complexity of the system, simulation-based planning environments are ideally suited for this. (Niemann 2007), (Gu 1997, Kimura 2000)

3.1.5.2 Operating data for monitoring systems

Desired times for handling orders are generated through preliminary costing. In the final costing, information needs to be constantly monitored and compared with real manufacturing data. The data required for this can be obtained from work plans and from PPS, PDE, MDE and QDE systems (Müller and Krämer 2001). Due to the hierarchical system structure, the manufacturing data can also be used to analyse sub-systems and even individual machines as well as to make comparisons with planning values from the preparatory work. In this way, concepts for improvement potentials can be identified for each level of a manufacturing system. The new values then form the basis for future planning cycles. Deviations from planning values supply information about fuzziness and model correction requirements. They also represent potentials in a manufacturing system which can still be realised by taking specific improvement measures.

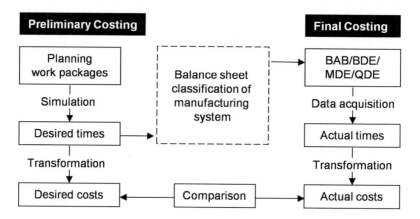

Fig. 3.12: Operating data as a basis for system monitoring. (Niemann 2007)

Figure 3.12 shows a continuous comparison for final costing as well as for correcting the forecast of time and cost data. The control model thus makes the progression of event, planning and evaluation data over time available. In this way, a control cycle can be created which constantly supplies updated references related to the planning cycles to be carried out. However, the improvements made to the system also need to be included in the work plans in order to realise the potentials in future planning cycles. The continuous planning, optimisation and controlling of manufacturing events thus enables the user to permanently control the efficiency of a manufacturing system throughout its entire life cycle. (Seliger et al. 1999), (Hieber and Schönsleben 2001), (Jackson et al. 1997) (Niemann 2007)

3.1.5.3 A reference model for manufacturing systems

First of all, system interrelationships are acquired and depicted using a reference model. In the process, the reference model should be filed with the control model for life cycle controlling to enable the depiction and evaluation of alternative manufacturing scenarios obtained using simulation. Performance units form the basis for this, via which the manufacturing transformation process takes place. The performance unit needs to be included in an information-related system environment within which continuous back coupling takes place with regard to the required performance (result).

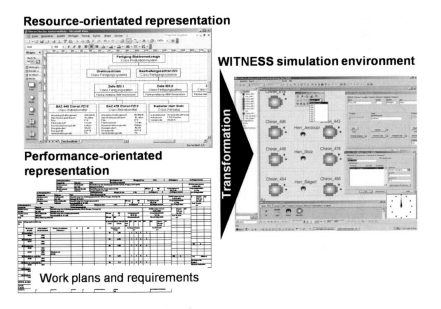

Fig. 3.13: Transformation of the manufacturing system into the simulation environment. (Niemann 2007)

By modelling the system in the markup language UML, it was possible to develop class diagrams in which the Stuttgart company model could be modelled completely with the individual hierarchical levels as well as formally described with regard to the corresponding usage characteristics and costs created. At each level, the classes contain specific and inherited attributes and functions with which they can be represented as far as their tasks and interaction with other classes in the manufacturing system are concerned. Additionally, through inheritance relationships, each individual class obtains all the properties of the class of performance units. (Niemann 2007)

In this way, a system of classes is formed with system structures and elements similar to their own at all levels of the system hierarchy. By precisely specifying the necessary data and also data sources in relation to the time and cost data required, the reference system developed can be coupled with the corresponding operating information system for acquiring and depicting real system conditions. As a result, a general model for manufacturing systems is created which can then be utilised as a basis for simulation-aided life cycle controlling (Figure 3.13). (Niemann 2007)

3.1.5.4 The control cycle for cost controlling

In order to safeguard the benefit of cost-intensive improvement measures in the long-term, a planning horizon needs to take a period of time as long as possible into consideration.

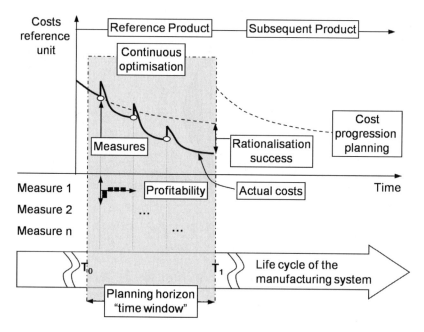

Fig. 3.14: Synchronising the product life cycle with the system life cycle. (Niemann 2007)

Depending on the degree of planning reliability and the volatility of the manufacturing program, this can also be selected directly right up to the intended final shutdown of the system. To do this, the corresponding paradigm of a life cycle-orientated calculation of manufacturing costs and the effect of potential improvement measures during the later life of the system need to be represented. Here, the short-, mid- and long-term effects have to be described and analysed. As planning cycles progress, the planning horizon (period of time) thus shifts continuously along the time axis in the direction of the intended shutdown of the system. (Niemann 2007) This is shown in Figure 3.14.

Due to the fact that the system is subjected to constant dynamics as a result of external and internal change drivers, the profitability of optimisation measures taken must be verified at regular intervals and maintained. The optimum operating point of a system has to be appropriately monitored and updated. To achieve this, the method needs to be transferred to a control model using which system configuration can be continuously adjusted to ensure optimum piece costs in dependence upon the orders being manufactured. (Niemann 2007)

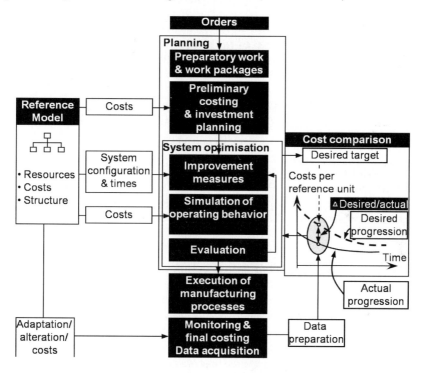

Fig. 3.15: Control cycle for life cycle controlling. (Niemann 2007)

As a result, product life cycles are synchronised with the existing manufacturing system for the duration of the planning time. This is because adjustments made to the system are always orientated towards the development of capacity packages. The controller is made up of both a planning and an optimisation environment in which the path of the manufacturing system's process sequences can be simulated.

In the controller (Figure 3.15), first of all capacitance packages are generated for the manufacturing system from the manufacturing orders. The packages represent the required total capacity of a specific, identical machine type.

Thus, using the work plans and the actual system configuration, the capacities desired can be calculated by adding up the main usage, set-up and technical loss times (Figure 3.16). These must then be compared with the nominal capacity of the capacitance package available.

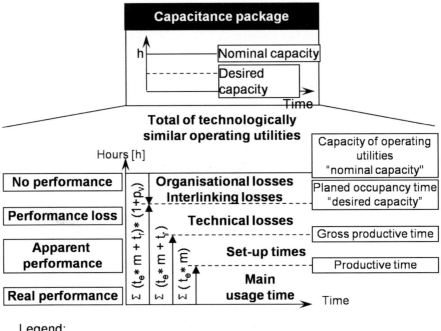

Legend:
t_e= processing time; m= number of components; t_r= set-up time;
p_v=technical losses

Fig. 3.16: Acquisition of required overall capacity using capacitance packages.

The actual nominal capacity is either already present in the reference model or is continuously imported and acquired via operating data administration. As part of the optimisation, alternative manufacturing scenarios or potential improvement measures can now be evaluated with regard to their effect on the life cycle (trade-offs) and on profitability. (Niemann 2003, 2007). As a result of developing alternative scenarios for the manufacturing system, more optimisation loops may need to be run in this area. For the analysis, these strategic options are first represented in the reference model and are then analysed using simulation.

The results obtained from the simulation give a chronological representation of the individual sequence types based on the current system configuration. The times and piece numbers ascertained form the basis for the cost assessment of system performance for the planning horizon under examination. For the approach selected, concrete operative measures are then developed with which the planned time savings in the system can be achieved.

The measures are then verified with regard to their financial requirement and also to their profitability and are then implemented in the manufacturing system in accordance with corporate management criteria. Manufacturing is carried out under the constant acquisition of operating data which are analysed in the final costing. As part of the controlling process, the times and costs forecast are compared with real system performance in order to monitor the achievement of objectives. The control cycle is closed by re-importing actual data from the final costing into the preparatory work and serves as a planning basis for future preliminary costing processes. Through the direct back coupling of the system planning with a time and cost analysis supported by operating data, a closed controlling system is created for a simulation-based control cycle with which manufacturing systems can be continuously planned, operated and optimised throughout their entire life cycle.

3.1.5.5 Industrial application of life cycle controlling

The method developed has been implemented in a practical application to control the life cycle of a manufacturing system used for the series production of machinable components.

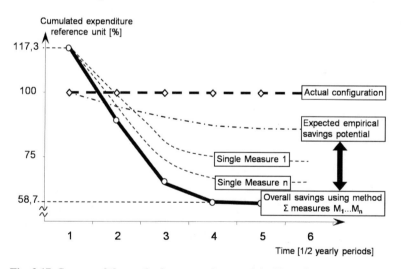

Fig. 3.17: Success of the method compared to empirical learning curves. (Niemann 2007)

By creating a control cycle of planned and actual manufacturing data, it was possible to identify rationalisation potentials and to evaluate them with regard to their profitability. In the process, it was demonstrated that the applied method enabled the company concerned to realise considerably-improved learning curves than could be expected using comparable empirical data (see Figure 3.17). The development of the ascertained cost pathways over time also forms a benchmark for the continuous controlling of system profitability. The increase in learning speed enables comparative cost advantages to be realised which contribute towards ensuring that company competitiveness is safeguarded on a lasting basis. (Niemann 2007)

It became possible to evaluate alternative manufacturing scenarios faster and to learn from the "future" by implementing measures virtually. By coupling planning data with actual data, a control loop of life cycle controlling was created which enabled manufacturing systems to be planned, optimised and monitored continuously throughout their life cycle. The constantly-updated system model thus formed the basis for future planning and optimisation cycles. The practical application of the controlling model has demonstrated that companies are able to realise lasting and effective rationalisation potentials as a result.

3.1.6 Life cycle cost contracts

More and more often, customers are demanding certainty regarding the follow-up costs of their investment goods. To do this, they frequently insist on manufacturers giving contractually-fixed guarantees concerning a large part of the maximum expected operating costs. These life cycle cost contracts (block guarantees) limit cost risks on the part of the customer and involve the plant manufacturer in product responsibility. At the moment, the huge demand for such delivery agreements is only met by a few concrete proposals on the part of the manufacturer.

A study carried out by the Institute of Industrial Manufacturing and Management (IFF) of the University of Stuttgart already shows today that "delivery alone" will no longer be sufficient in the future to establish lasting business relationships with customers. (Niemann,J. and Stierle, T. 2004) However, the internal path towards a manufacturer supplying such "assured" offers is long. To be able to do this, manufacturers need to know the operational behaviour of their products precisely in order to be capable of giving customers a life cycle cost guarantee. Otherwise, considerable risks due to possible penalties could result. The contracts also obligate the system operator to comply with the operating conditions laid down by the manufacturer (e.g. fixed maintenance cycles, etc.). With the aid of an example, Figure 3.18 shows a basic method for designing a life cycle cost contract between the manufacturer and the customer.

The benefit for both partners lies in the exploitation of these synergies. Examples already show that companies consistently implementing this model as a life cycle orientated partnership are able to operate with strategic competitive advantages in the market.

The IFF study also demonstrated that there is a particular lack in services, thus enabling the forceful utilisation of machine performance potentials. (Niemann,J. and Stierle, T. 2004) The potentials here range from disturbance management right up to the preparation of specialised manufacturing know-how. In the future, innovative payment models will also play a role which will be orientated towards the benefits brought by the machine supplied.

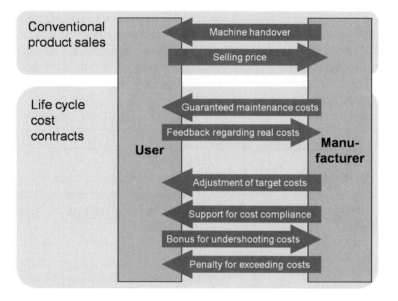

Fig. 3.18: Example of a life cycle cost contract.

In the process, conventional machine sales will be extended by business models using which the machine manufacturer will be paid for the machine performance utilised, ie. the benefit (re-) sold. As a result, the life cycle benefit of the investment becomes a focal point once again and it is in the mutual interest of both the machine manufacturer and the user to optimise this from the point of view of maximising production yield. This especially leads to closer collaboration between both partners in which the conventional customer-supplier relationship changes to become a cooperative system partnership.

3.2 Ecological evaluation

Life Cycle Assessment (LCA) is a concept for the evaluation of environmental performance. It is a method for acquiring and evaluating the effects of products, processes and different types of services on the environment over their complete life cycle from the mining of raw materials and usage of a product right up to its disposal. Weaknesses, both ecological and economical, can be identified by way of life cycle assessment. This comparison is carried out in accordance with the current international norm for evaluating environmental performance (DIN ISO 14040 ff). At the same time, it also enables a basis to be created for the informed communication of environmental protection successes which are related to the product. The use of this method is described in more detail by way of an example in the following:

3.2.1 The application of life cycle assessment

The production of high-precision steel parts usually includes a hardening process for altering the surface structure. Conventional heat treatment methods are characterised by high energy consumption and the utilisation of polluting treatment salts.

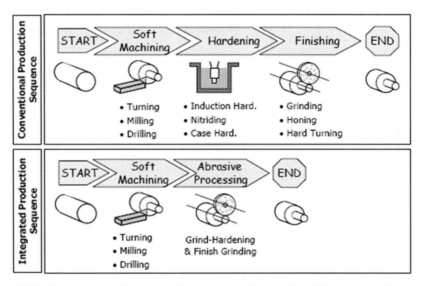

Fig. 3.19: Comparison of a conventional production chain with a production chain including grind hardening.

Grind-hardening is an alternative process which can be used for the simultaneous surface hardening and grinding of metallic components. In a study by Salonitis et. al., life cycle assessment is used for the environmental analysis of the grind-hardening process (Salonitis et al. 2006). In two different pilot cases - the production of raceways and the production of tripod joints - the environmental impact of the grind-hardening process is compared with that of conventional heat treatment methods:. The analysis indicates that the utilisation of grind-hardening significantly decreases the environmental impact. Additionally, the heat treatment process is usually performed by external subcontractors. This involves transporting the workpieces back and forth, further increasing both energy consumption and the environmental impact. The necessary cleaning of the workpieces before and after the heat treatment process also requires copious amounts of water. All the above-named steps can be eliminated by performing both the grinding and grind-hardening processes on the same machine with the same set-up (Figure 3.19).

3.2.2 Further studies on ecological assessment

The environmental assessment (EA) of a process includes evaluation of the consequences for the environment, human health and resources. It is actually a part of a product's life cycle assessment. In the study of Drakopoulos et al., the product is a ship and the specific processes refer to the repair phase of the ship's life cycle (Drakopoulos et al. 2006). In the case of ship repair, where it is evident that the most dangerous environmental processes are cutting and welding, the goal of EA has been to quantify this danger and to benchmark the various different processes used. This study presents an environmental analysis of a number of cutting and joining processes carried out during the repair of a ship's hull. These processes include *oxyacetylene cutting*, *plasma arc cutting*, *shielded metal arc welding*, *flux core arc welding* and *submerged arc welding* and are modelled in terms of their environmental impact. The environmentally-related inputs and outputs of each process are elaborated with a life cycle assessment tool. The impact on various aspects, such as human health, resource depletion etc., is evaluated by the impact assessment methods of "Environmental Priority Strategy (EPS)" and "Eco-Indicator 99". Based on the results, the cutting and welding processes are benchmarked in terms of their environmental impact.

3.3 Interim summary of life cycle evaluation

The chapter has demonstrated that the sustainable design of product life cycles must be coupled with a continuous assessment of the economical and ecological consequences. In order to do this, standardised methods are required for estimating the effects occurring during the life cycle. Here, a qualified evaluation is based on the generation of corresponding life cycle situation-related models (see Chapter 2). However, the methods known to date in research and industrial practice also show that a once-only calculation at the start of production development is inadequate. This is because there is a significant probability that changes will take place during the long life cycle of the product, thus making the original planning measures obsolete. This means that especially cost-orientated planning scenarios will always be flawed by considerable uncertainty. The only remedy is to introduce continuous planning systems which are constantly adjusted over the life cycle in the form of life cycle controlling to support decision-making at each of the planning horizons. The case study shows that a continuous planning approach contains considerable potentials because it enables the user to constantly optimise the product life cycle by "learning from the future".

3.4 References concerning chapter 3

(Blanchard 1978) Blanchard, B.: Design and manage to life cycle cost, Portland, Or.: M/A Pr., 1978.

(Brussel and Valckenaers 1999) Brussel, H. van, Valckenaers, P. (Hrsg.): Katholieke Universiteit <Leuven> / Department of Mechanical Engineering / Production Engineering Machine Design Automation (PMA): Intelligent Manufacturing Systems 1999: Proceedings of the Second International Workshop on Intelligent Manufacturing Systems, September 22-24, 1999, Leuven, Belgium.

(Drakopoulus et al. 2006) Drakopoulos, S., K. Salonitis, G. Tsoukantas and G. Chryssolouris, 2006, "Environmental Impact of Ship Hull Repair", Proceedings of the 13th CIRP International Conference on Life Cycle Engineerin , Leuven, Belgium, pp. 459-464.

(Eversheim and Schuh 1999) Eversheim, W. (Hrsg.); Schuh, Günther (Hrsg.): Gestaltung von Produktionssystemen. Berlin u.a.: Springer, 1999 (Produktion und Management 3)

(Gu et al. 1997) Gu, P.; Hashemian, M.; Sosale, S. (1997): An integrated modular design methodology for life-cycle engineering. In: Annals of the CIRP, 46(1), S. 71-74

(Hieber and Schönsleben 2001) Hieber, R. (Hrsg.); Schönsleben, P. (Hrsg.) (2001): Supply Chain Management: A Collaborative Performance Measurement Approach. Zürich: vdf-Hochschulverlag AG an der ETH Zürich

(Hummel and Westkämper 2006) Hummel, Vera; Westkämper, Engelbert (2006): The Stuttgart Enterprise Model - Integrated Engineering of Strategic and Operational Functions. In: Manufacturing Systems 35, Nr. 1, S. 89-93

(Jackson et al. 1997) Jackson, P.; Wallace, D.; Kegg, R. (1997): An analytical method for integrating environmental and traditional design considerations. In: Annals of the CIRP, 46(1), S. 355-360

(Kimura 2000) Kimura, F. (2000): A Methodology for Design and Management of Product Life Cycle Adapted to Product Usage Modes. The 33rd CIRP International Seminar on Manufacturing Systems, 5-7 June 2000, Stockholm, Sweden

(Müller and Krämer 2001) Müller, P.; Krämer, K.; REFA Verband für Arbeitsgestaltung, Betriebsorganisation und Unternehmensentwicklung: BDE mDE Ident Report 2001: Das Handbuch zur Datenerfassung mit Marktübersicht, Anwenderberichten, Anbieterverzeichnis und Auswahlsoftware; Betriebsdatenerfassung, mobile Datenerfassung, Ident-Techniken (Auto-ID). Darmstadt, 2001 (Edition FB/IE)

(Niemann 2003a) Niemann, J. (2003): Ökonomische Bewertung von Produktlebensläufen-Life Cycle Controlling. In: Bullinger, Hans-Jörg (Hrsg.) u.a.: Neue Organisationsformen im Unternehmen : Ein Handbuch für das moderne Management. Berlin u.a. : Springer, p. 904-916

(Niemann 2003b) Niemann, J.: Life Cycle Management, In: Neue Organisationsformen im Unternehmen - Ein Handbuch für das moderne Management, Bullinger, H.-J., Warnecke, H. J., Westkämper E. (Ed.), 2. Auflage, Springer Verlag, Berlin u. a.; 2003

(Niemann 2007) Niemann, J. 2007. Eine Methodik zum dynamischen Life Cycle Controlling von *Produktionssystemen*. Stuttgart, Germany: University of Stuttgart (Dissertation). Heimsheim, Germany: Jost-Jetter.

(Niemann et al. 2004) Niemann, Jörg; Stierle, Thomas; Westkämper, Engelbert: Kooperative Fertigungsstrukturen im Umfeld des Werkzeugmaschinenbaus : Ergebnisse einer empirischen Studie. In: Wt Werkstattstechnik 94 (2004), Nr. 10, S. 537-543

(Niemann and Westkämper 2005) Niemann, J., Westkämper, E. (2005) : Dynamic Life Cycle Control of Integrated Manufacturing Systems using Planning Processes Based on Experience Curves. In: Weingärtner, Lindolfo (Chairman) u.a., CIRP: 38th International Seminar on Manufacturing Systems / CD-ROM: Proceedings, May 16/18 - 2005, Florianopolis, Brazil. p. 4

(Salonitis et al. 2006) Salonitis, K., G. Tsoukantas , S. Drakopoulos , P. Stavropoulos and G. Chryssolouris, 2006. "Environmental Impact Assessment of Grind-Hardening Process", Proceedings of the 13th CIRP International Conference on Life Cycle Engineering, Leuven, Belgium, pp. 657-662.

(Seliger et al. 1999) Seliger, G.; Grudzien, W.; Zaidi, H. (1999): New Methods of Product Data Provision for a simplified Dissassembly. In: Proceedings of the Life Cycle Design 99, Kingston, Kanada

(Spath 2003) Spath, D. (ed.) (2003): Ganzheitlich produzieren. Innovative Organisation und Führung. Stuttgart: LOG_X Verlag GmbH.

(Statistisches Bundesamt 2007) Statistisches Bundesamt: Statistisches Jahrbuch 2004 für die Bundesrepublik Deutschland. Wiesbaden, 2007

(VDMA 2006) VDMA-Verein Deutscher Maschinen- und Anlagenbau e.V. 2006, VDMA 34160: Forecasting Model for Lifecycle Costs of Machines and Plants. Berlin, Germany: Beuth Verlag.

(Warnecke 1993) Warnecke, H.-J.: The Fractal Company - A Revolution in Corporate Culture. Berlin u.a.: Springer, 1993

(Westkämper 2003) Westkämper, E. (2003), Wandlungsfähige Organisation und Fertigung in dynamischen Umfeldern, in: Bullinger, H.-J., Warnecke, H. J., Westkämper, E. (Hrsg.): Neue Organisationsformen im Unternehmen - Ein Handbuch für das moderne Management, 2. neu bearb. und erw. Auflage, Berlin u.a.: Springer Verlag

(Westkämper 2004) Westkämper, E. (Leitung); Drexler, K. (Red.); Moisan, A. (Red.); CIRP: Wörterbuch der Fertigungstechnik-Band 3: Produktionssysteme. 1. Aufl.. Berlin u.a.: Springer, 2004

(Westkämper 2004) Westkämper, E.: Das Stuttgarter Unternehmensmodell: Ansatzpunkte für eine Neuorientierung des Industrial Engineering. In: REFA Landesverband Baden-Württemberg: Ratiodesign: Wertschöpfung - gestalten, planen und steuern. Bodensee-Forum. 17. und 18. Juni 2004, Friedrichshafen. Mannheim, 2004, S. 6-18

(Westkämper 2006) Westkämper, E.: Einführung in die Organisation der Produktion. Strategien der Produktion. Berlin; Heidelberg: Springer, 2006

(Wübbenhorst, 1984) Wübbenhorst, K.: Konzept der Lebenslaufkosten, Darmstadt, Verlag für Fachliteratur, 1984.

(Wübbenhorst 1992) Wübbenhorst, K. L.: Lebenszykluskosten, in: Schulte, C. (Hrsg.), Effektives Kostenmanagement, Methoden und Implementierung, Schäffer-Poeschel-Verlag, Stuttgart 1992, S. 245-271.

4 Life cycle information support

In the preceding chapters, methods for modelling product life cycles and for assessing and optimising the associated economical and ecological effects were presented. However, in order to model and evaluate, data regarding events taking place in the various phases of the product life cycle are required. The following chapter is concerned with finding ways to effectively support and optimise the sustainable design of product life cycles through the acquisition and analysis of life cycle data. (Figure 4.1)

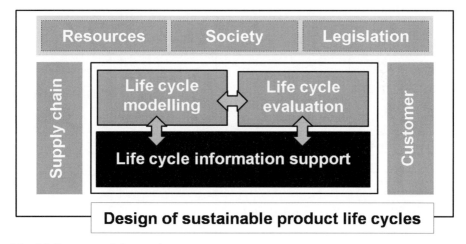

Fig. 4.1: Structure of chapter 4.

First of all, the important aspect of the continuous acquisition and control of the huge quantities of data is considered. Here, the relevant sources of data need to be identified both in the product design phase and especially during product usage in order to enable product life cycles to be adequately recorded and tracked. These data have to be constantly processed so that they can be appropriately utilised according to the relevant situation and product usage by the corresponding life cycle partner as part of holistic product management. Only then is it possible to continuously optimise product usage over time in the fringe ranges of performance and precision.

4.1 Reliable data for transparent product life cycles

The critical factor for success in these developments is the management of data and information. The volume of information is exploding and industry needs actual and reliable information on the state of the art. Life cycle management provides the opportunity to maximise the benefits of each product in all its phases. To do this, life-long information which is continuously updated is required.

A future development has to take into account every option of implementing all basic data of products into their internal information system. This would help to support all product-related operations and associated activities with actual documentation. (Eversheim and Schuh 1999), (Müller and Krämer 2001), (Niemann 2003), (Warnecke 1993), (Westkämper 2003), (Westkämper 2004), (Westkämper et al. 2004)

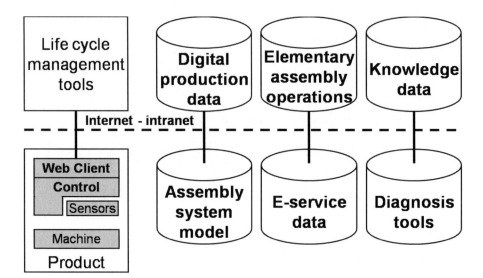

Fig. 4.2: Platform for integrated product life cycle management.

Future life cycle management systems are open systems which operate using communication standards and permit the implementation of product- or customer-specific IT applications. Figure 4.2 summarises such a platform which contains basic communication functions as well as specific systems for supporting products and operations with data during all phases of life. Such platforms are now on the market for supporting e-services. (Niemann 2005), (Niemann and Westkämper 2005)

Basic standards for managing and exchanging data are also available. Key problems here include the updating of data and the protection of know-how.

In the automobile industry, worldwide standards have been developed for the exchange of life cycle data and for the management of services, also including logistics. For the machine industry, it would make sense to start by developing basic product and process models to find solutions both for standardisation as well as for the application of new services during the life of technical products.

4.2 Digital product tracking

Today's traditional product business models see a product in the centre of all activities which is surrounded by peripheral services. These (additional) services are either provided by the supplier or by other service agents (third parties). Such business models are currently changing towards so-called participative or even full-service concepts. (functional sales, product service systems) This implies that the supplier extends (or sometimes has to extend…) his operations over the product's life cycle. Modern business models require a supplier guarantee of system cost per piece. Alternatively, the customer only pays for the usage of a machine rather than for the purchase of the machine itself. This results in a dramatic change in business relationships between users and suppliers because the original equipment manufacturers (OEMs) transfer the production risks to their suppliers. The supplier is responsible for the operation of a machine and is paid according to output.

However, in many industries suppliers often possess considerably more in-depth knowledge about machine operation than the OEM. Consequently, they are able to operate their machines in the fringe ranges of performance and quality. They benefit from their long time experience and in-depth constructional knowledge which enables them to improve machine reliability and availability in order to achieve improved productivity (resource efficiency) compared to that of a machine operated by the OEM. Another stimulus to follow these new business relationships is the evaluation of a customer's value over the entire business relationship (customer lifetime value). Modern business models create extensive and long-lasting business partnerships (Figure 4.3). Such durability is achieved by integrating a partner right from the planning phase of production systems. This allows the supplier to generate pricing models according to the degree of service requested. The most extensive service package is total system performance management (full-service provider). If the customer chooses this option, the supplier then takes over total control of all production operations as well as complete responsibility for the product. Thus, the supplier mutates into a system operator with a strong influence on OEM performance (and profit).

This degree of integration makes it both difficult and costly to replace such a partner who was involved in the essential planning and scaling of the system. The system, the tasks to be performed, logistic chains and system management were all designed to achieve a maximum profit among all partners. Also, the knowledge and abilities of all the partners were distributed according to the working tasks, making it almost impossible to find another partner with exactly the same special knowledge profile.

The close integration of suppliers permits OEMs to minimise their production risks. However, at the same time, production costs change from being fixed to becoming variable parameters. On the other hand, suppliers are able to extend their value chain and apply their special machine expertise to realise additional potentials. This constitutes the basis for a durable win-win relationship for both partners.

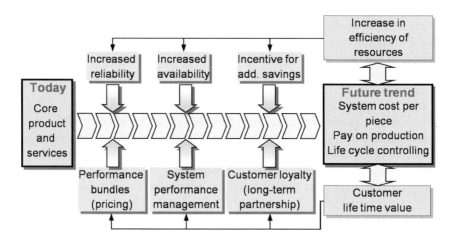

Fig. 4.3: Performance potentials for system operators.

When implementing this strategy, it becomes obvious that a supplier who has to constantly guarantee life cycle or production system costs also needs online access to the system control. The supplier has to permanently monitor his machines and be capable of reacting at short notice in case of failures, machine breakdowns or deviations in quality. Any irregular events directly worsen his profit situation. Most performance contracts (e.g. life cycle cost contracts) include penalties if the fixed performance criteria are not met but also monetary motivations for outstanding performance results over and above fixed ratios.

Digital lifetime tracking and (online) optimisation are therefore essential in order to meet the fixed performance criteria (Figure 4.4). A general model has to cover the entire machine life cycle, starting with the design of the machine and ending its "death" through recycling. Similar to a patient's file at a doctor's practice, this digital machine file can be considered as a document where all machine data and events have been logged. In accordance with the life cycle, the bill of materials (design data), dates of sale, inspections, changed parts (usage data) and recycled parts (recycling data) are all recorded continuously.

The performance or service contracts demand online tele-operations and short reaction times in order to optimise operations in case of abnormal machine behaviour. A direct look at the data and current performance parameters of the machine enables financial ratios to be constantly generated. The values can be used for life cycle-orientated machine controlling, including pro-active performance and profit management based on technical performance control.

Today, the Internet offers a wide range of usable tools for life cycle management and there are worldwide standards for the communication and exchange of data and information. The Internet uses tools, engines and robots (behind the interfaces) to search for information and knowledge. There are new standards for b2b (business to business) and b2c (business to consumer) communications. Leading companies in the automotive and machine industries use the Internet as a platform for logistics and for administrating processes between OEM and suppliers.

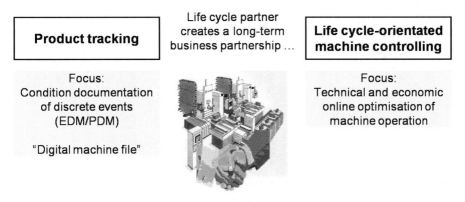

Fig. 4.4: Digital life cycle product data management.

Internet technologies offer a broad spectrum of tools for managing the link between manufacturers and users wherever they are located. Examples of this include the management of component supply logistics for assembly and maintenance or technical support in the usage phase. The basic architecture of the Internet and of intranets in company information technologies is illustrated in Figure 4.5. It shows three systems for internal, service partner and common information.

Security techniques (firewall, en- or decryption) have to be adapted to the needs of manufacturers and be able to be operated in closed areas with service partners. Both the architecture of a product's control systems and Internet availability at each work place are essential for assembly and maintenance.

Functional diagnosis and usage monitoring can be integrated into these systems and linked to the Internet. The same diagnosis systems are required both for final assembly and for maintenance. New internal information system architectures follow agents' theories. The development of Product Embedded Information Devices (PEID) is expected to progress rapidly. It will be largely used for advanced product life cycle management and real-time data monitoring throughout the product supply chain. This technology will especially allow producers to dramatically increase their ability and capacity to offer high-quality after-sales services and will also enable them to demonstrate their fulfilment of responsibility as producers of environmentally friendly and sustainable products.

In general, the product life cycle consists of three main phases: beginning of life (BOL), which includes design and production; middle of life (MOL), which includes logistics (distribution), use, service and maintenance; and end of life (EOL), which includes reverse logistics (collecting), remanufacturing (disassembly, refurbishment, reassembly, etc.), reuse, recycling and disposal. The information flow is quite complete during BOL because it is supported by several information systems such as CAD/CAM, product data management (PDM) and knowledge management (KM).

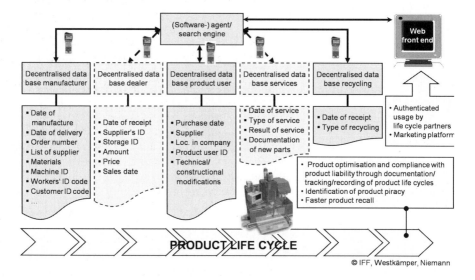

Fig. 4.5: Traceability of products during an entire life cycle.

However, the information flow becomes vague or unrecognisable after BOL. This prevents the feedback of data, information and knowledge from MOL and EOL back to BOL. As a result, the life cycle activities of the MOL and EOL phases have a limited visibility of product-related information. Consequently, actors involved in each life cycle phase make decisions based on incomplete and inaccurate data, resulting in operational inefficiencies (IMTI 2002).

However, over the last decade, a rapid development of Internet, wireless mobile telecommunication technologies and several product identification technologies (see Table 4.1) have changed our stereotype of the product life cycle. These technologies allow us to visualise product information throughout the entire product life cycle. The core element of these technologies is the product identification mechanism which has been given several different names; *smart tag* (Qiu and Zhang 2003), *Auto-ID* (Udoka 1991, Parlikad *et al.* 2003) or *intelligent product* (Wong *et al.* 2002, McFarlane *et al.* 2003).

		Product life cycle		
Technologies	Definition	BOL	MOL	EOL
Auto-ID: EPC (Parlikad et al. 2003)	Electronic Product Code: Product unique code	•	•	•
Auto-ID: PML (Brock et al. 2001)	Physical Markup Language: Markup language for product information	•	•	•
Auto-ID: ONS (Foley 1999)	Object Naming Service: Telling computer systems location information on the Internet about any object that carries an EPC.	•	•	•
ID@URI (Huvio et al. 2002)	Identifying physical product items and linking to the product agents that handle their information	•	•	•
RFID (Schneider 2003)	Radio Frequency IDentification: Communication technology for collecting and transferring information via radio waves	•	•	•
GPS (Evers and Kasties 1994)	Global Positioning Systems: Satellite navigation system used for determining one's precise location and providing a highly accurate time reference	–	•	•
GIS (Evers and Kasties 1994)	Geographical Information System: Information system capable of assembling, storing, manipulating and displaying geographically-referenced information	–	•	•

Table 4.1: Product identification technologies.

During the product life cycle, a great deal of information related to the product is generated, ranging from business meeting notes and marketing requirements to CAD drawings, test results, repair instructions and right up to sales reports. The quality of a product and the processes during the product life cycle can be improved by enhancing the traceability of this product life cycle data. Most of the data can be acquired during production, distribution and procurement of the components as well as during the manufacture and maintenance of the product. In order to support the extensive traceability of an individual product throughout its entire product life cycle, each life cycle actor needs to be provided with all the required information by tracing the product life cycle data with the product identification technologies mentioned in Table 4.1.

4.3 Boosting utilisation performance

4.3.1 The phase of product design

The quality and shelf life of today's technical products is no longer determined by wear and attrition but by being technically out of date. Modern strategies for maximum product utilisation have to consider long-term lifetime planning for the product. Life cycle management therefore aims to achieve maximum product performance throughout the entire life span covering the phases of design, usage and recycling. The planning also has to take into consideration the needs of all partners during the product's lifetime. (Westkämper 2006) The traditional focus on optimising partial processes only reaches a sub-optimum in the value-added chain. Potentials offered by synergies and trade-offs can only be activated and realised by coordinating all of the phases of life (Figure 4.6).

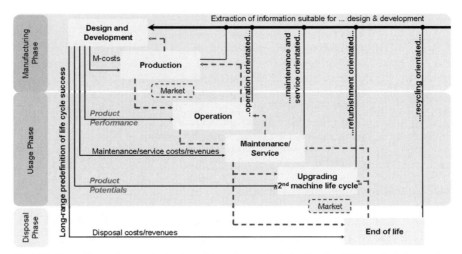

Fig. 4.6: Overall product optimisation through constant feedback of life cycle information. (fig. enhanced acc. to (Eversheim and Schuh 1999))

In this respect, the design phase is a key driver for sustainable lifetime success. This is because not only product functions but also product life cycle costs and revenues are all determined in this early phase. (Westkämper 2003) (Westkämper et al. 2004).

Design parameters determine product performance as well as future product performance potentials (2nd product life) which may be activated in later phases of life. The option of upgrading a product (2nd product life) offers additional potentials to optimise the overall product through the constant feedback of life cycle information. Therefore, the design phase has a deep impact on total life cycle performance.

4.3.2 The phase of product utilisation

Today's new machine control concepts provide access to machine data. The programmable logic controls (PLC) generally used are increasingly set up in a modular fashion and enable flexible application. Combined with intelligent machine and field buses, they allow the realisation of fractal machine control systems. Similar to nervous systems, the control tasks are distributed among more central components such as master computers as well as among more decentralised components right down to the actor/sensor level. (Niemann et al. 2001), Xie et al. 2002), (Cantamessa and Valentini 2002), (Rehmann und Guenov 1998)

This concept is supported by the development trend towards PC-based controls. Utilisation of these life cycle database improves the performance of distributed artefacts in terms of reliability, availability, maintainability and serviceability. "Transparent" machine control over a long distance is practically state-of-the-art even though only a minority of machine manufacturers use this technology to provide customer support. A multitude of functions, such as putting a machine into operation, carrying out maintenance and even actually operating a machine can be supported this way. (Pritschow et al. 1998), (Berger et al. 1998)

There are a number of applications for remote machine control and service which use telecommunication: the most common application is the access to the control software, e. g. for the purpose of analysis, error diagnosis or updating. Other applications result from the compression, transmission and evaluation of sensor data, e. g., for condition monitoring where sensors enable the mechanical wear of component parts to be monitored. The most recent approaches focus on establishing life cycle databases and tracking machine behaviour and performance data.

Figure 4.7 shows a structure of the network of services for a machine. The network is characterised by connections which allow the transfer of knowledge and information both automatically as well as manually. The nodes serve as a provider, server and distributor of knowledge. In this way, complex structures are generated which consist of knowledge sinks and sources and in which communication via the web is made possible using transparent interfaces.

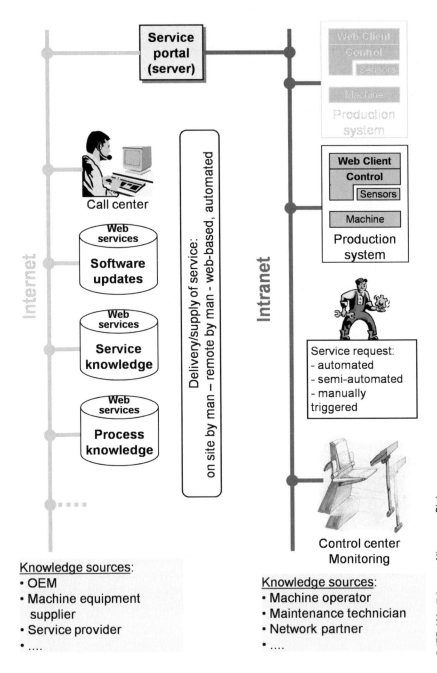

Fig. 4.7: Knowledge sources and sinks for e-service.

4.3.2.1 Permanent performance evaluation

Most of these data are generated during the production phases right up to the end of the ramp-up phase. In the usage phases, especially the maintenance phase, the basic data concerning the actual condition of the product changes constantly. Service, diagnosis and disassembly need data from current operations and planned activities in the same way as in the case of assembly.

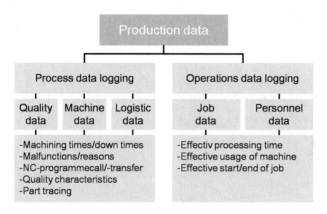

Fig. 4.8: Framework for manufacturing data acquisition.

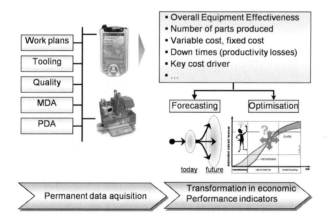

Fig. 4.9: Economic machine performance control.

By monitoring the actual process statuses (process monitoring) and compressing and comparing real data with planned data (process control), it is possible to evaluate process quality. Process information systems provide information support for this task. At the moment, the main challenges involved here include increasing process orientation and implementing the necessary organisational and information tasks.The relevant processes which contribute largely towards the result need to be selected from the wide range of company processes involved. In order to obtain sound process control information, the real data occurring needs to be compared with the planned data. (Figure 4.8 and 4.9) Process information systems are much more than process cost calculation systems assisted by data-processing. This is because data are determined which are mainly concerned with value rather than quantifiable data such as cycle times or the time factor of a process organisational unit. (Niemann 2007)

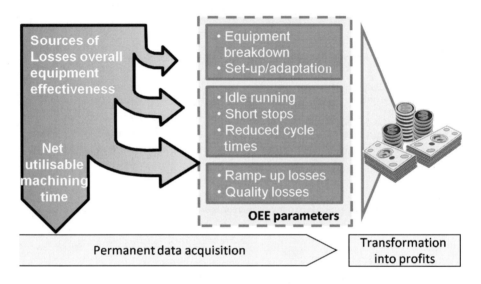

Fig. 4.10: Overall equipment effectiveness (OEE).

Permanent machine data acquisition also allows overall equipment effectiveness (OEE) to be evaluated. This measurement has its origin in the philosophy of total productive management. The OEE measures all losses occurring during machine operation. All sources of loss are accumulated into one % factor, the OEE factor. This number ranges from 0 to 1 meaning that a factor of 1 (or 100%) describes the optimal machine performance. The OEE factor consists of 3 major loss categories (Figure 4.10). The data required to analyse and aggregate these categories can mainly be acquired from machine and production operation. As a result, machine costs can be analysed and optimised using these control systems. The losses identified by the OEE analysis can be interpreted as lost profits (or additional cost, *opportunity cost*) because parts cannot be sold due to inappropriate machine operation and additional labour and material costs result. The equipment effectiveness losses measured can be expressed as a coefficient and transformed into economical values by linking them to resource process cost rates. The expression of this *performance loss* usually represents an enormous and often-underestimated value-adding potential in production. The retrograde and cumulated analysis of production data also shows up the main cost drivers and "expensive" work steps. This knowledge is useful for reorganisation planning, re-engineering and long-term technology planning purposes. (Niemann 2007)

4.3.2.2 Improvements on current products

In an environment characterised by hard competition to gain market shares, companies are striving to find new ways of providing additional value to customers and giving them a competitive edge over their competitors. Total management of the product life cycle is critical in order to innovatively meet customer needs throughout the entire life cycle. This is especially true during the usage phase and must be accomplished without increasing costs, sacrificing quality or delaying product delivery. The range of such holistic products and supporting services available is currently limited by the information gap in the product's life cycle (Kiritsis et al. 2003). The closure of the information loops within and between product life cycle phases was one of the main motivations for starting PROMISE (Product Life cycle Management and Information Tracking using Smart Embedded Systems) project.

Continuous monitoring of the status and condition of a system or a component by using appropriate product-embedded information devices (PEIDs) provides a means for determining the right time interval for realising maintenance operations. It also assists in deciding about the type of maintenance operation and the individual or team to do it. This allows higher product availability and improved function during the usage phase.

The parameters which should be measured to monitor the status and condition depend on the type of product concerned. For example:

- For a truck, the main parameters to be measured are the oil temperature, engine rpm, engine load, etc.,
- For machines, the main parameters to be measured are the axis response to a velocity, temperature inside the working area of the machine, level of vibration, etc.,
- For refrigerators, the main parameters to be measured are the internal temperature, external temperature, compressor time on and off, etc.

The processing of this type of real time field data (real time field data are filed data verified at sufficiently small intervals) often requires diagnostic and prognostic processes.

The main purpose of the diagnostics module is to assess the current state and operational condition of the critical components in the system using on-line sensor measurements (Vachtsevanos and Wang 1999). This can be achieved through the continuous measurement of the values of a set of appropriate parameters. A comparison of the diagnostics approaches and their strengths and weaknesses is given in Table 4.2 (Schroer 2002).

Approach	Strengths	Weaknesses
Rule-based	Easy to understand due to their intuitive simplicity	Development and maintenance can be long and time-consuming
	Well-proven, with many deployed applications.	Knowledge acquisition bottleneck.
	Inference sequence can be easily traced.	Generally only faults anticipated during the design phase can be diagnosed
Models based on structure and behaviour	As the model is a "correct" model, theoretically all faults can be diagnosed. However, in practice this is difficult to achieve.	Computationally intractable on models with large numbers of components.
	With the appropriate software it should be possible to generate models from CAD data.	Generating adequate behavioural models for complex devices (e.g. a microprocessor) is a serious challenge. Developing a complete and consistent model is difficult. For example, a "correct" model will not be able to diagnose a bridging fault.
Diagnostic inference models	Provides good diagnoses when good sources of diagnostic information are available	Only usable where good sources of diagnostic information are available, therefore diagnostic issues will have to be considered at design time.
Case-based	A fairly intuitive and easy to understand process	Can only diagnose once an adequate case base has been built.
	The knowledge acquisition bottleneck can be overcome as learning is continuous and incremental.	It is not always apparent how diagnostic inferences are arrived at.
Fuzzy logic and neural networks	Good at dealing with incomplete and inaccurate information	As a sole approach, their ability to diagnose complex systems is questionable. However, combined with other approaches, a good solution may be possible.

Table 4.2: Comparison of diagnostics approaches. (Schroer 2002)

The main purpose of the prognostics module is to analyse the input from the diagnostics module as well as historical field data using appropriate models in order to draw a picture of the current situation and indicate potential consequences for the future (Vachtsevanos and Wang 1999).

Appropriate decisions can be made based on the results of the prognostics process regarding the future behaviour/condition of the component or system.

There are numerous benefits in recording filed data through the continuous monitoring of product status and condition which contribute towards providing additional value to customers e.g. in terms of service and maintenance. However, several problems still need to be further investigated:

- Privacy issues: does the manufacturer have the right to monitor his products during usage after they have been delivered to customers?
- Technical issues: do the current PEIDs have enough capability to record all the required data with the right quality and in the right quantity?
- Research issues: are all the parameters necessary for monitoring the condition/status of a system or component known and measurable?

4.3.2.3 Improvements on subsequent product generations

Regular improvements in design, especially through the use of knowledge generated from product field data, are essential in order to provide customers with products which match their requirements better. According to Oh and Bai (Oh and Bai 2001), field data are superior to laboratory data because they capture actual usage profiles as well as combined environmental exposures which are difficult to simulate in the laboratory. Coit and Dey indicate that even the most faithful and rigorous laboratory testing will fail to precisely simulate all field conditions (Coit and Dey 1999). Therefore, the principal advantage of using field data is the fact that actual operational and environmental stresses are taken into consideration. In the field, the stresses are applied simultaneously and variable interactions are implicitly considered by any analysis using this data. The enormous progress made in the development of sensors and other product-embedded information devices (PEIDs) such as Radio Frequency Identification tags (RFIDs) and in measurement technology enables field data about various aspects of the product such as reliability, availability and maintainability (RAM), life cycle costs (LCC), safety, environment, etc. to be acquired.

The transformation and analysis of product field data can provide valuable knowledge which assists designers in improving specific aspects of their (re)design activities such as RAM/LCC, safety and environment. To make this possible, the data need to be recorded with a focus on these aspects.

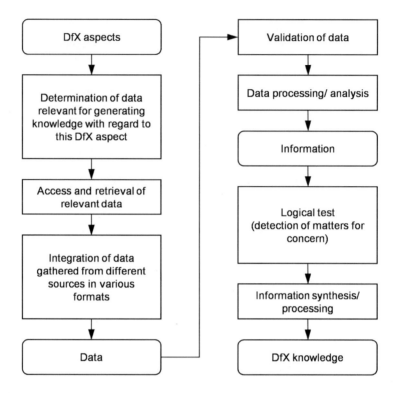

Fig. 4.11: Main steps in transforming product field data into knowledge.

The main steps involved in the process of transforming product field data into DfX knowledge are shown in Figure 4.11.

4.3.2.4 Impacts of RAM aspects

Improvements in reliability and maintainability through the use of design for maintainability and design for reliability knowledge, increases the availability (Figure 4.12) of products and has the following benefits:

- lower maintenance costs related to labour, material and logistics,
- increased user satisfaction by improving product availability. This is achieved by reducing the number of serious incidents requiring maintenance,
- improved company marketing image by emphasising the availability performance of the products. This can favour keeping long-term relationships with current customers as their availability problems are taken into account.

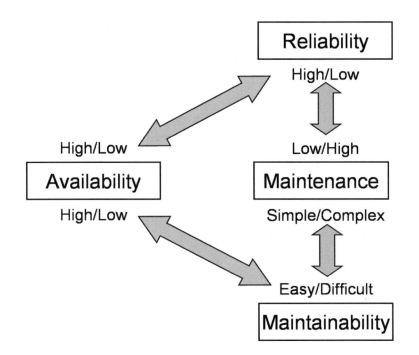

Fig. 4.12: Impact of improving reliability and maintainability on availability.

4.3.2.5 Impacts of safety aspects

According to Dowlatshahi there are four factors which affect the improvement in the safety of a product: (i) product liability litigation, (ii) governmental regulations, (iii) internal safety efforts and (iv) marketing requirements (Dowlatshahi 2001). The first two mentioned are *forced factors* and the third and fourth ones are *voluntary factors*. Consequently, improvements in product safety through the use of design for safety knowledge may have the following benefits:
- Increased user satisfaction by ensuring a safer usage environment,
- Reduced number of liability litigation problems and associated costs,
- Reinforced compliance with governmental regulations regarding safety requirements,
- Improved company marketing image by emphasising the safety performance of the products manufactured. This can favour keeping long-term relationships with current customers as their safety problems are taken into account.

4.3.2.6 Impacts of environmental aspects

Improvements in the environmental performance of a product through the use of design for environment knowledge may have the following benefits:
- Reduced costs by decreasing energy consumption and selecting efficient EOL alternatives for treating retired products,
- Reinforced compliance with governmental regulations and EU directives through environmentally friendly treatment of EOL products, reduced environmental burden of the usage phase and implementation of new material restrictions,
- Improved company marketing image by emphasising the environmental performance of the products manufactured. This can favour keeping long-term relationships with current customers as their environmental problems are taken into account.

4.3.3 The phase of upgrading and recycling

Maximum utilisation strategies include a future-orientated life cycle plan for the product. On completion of the usage phase, the owner is faced with the alternative of either scrapping/ recycling the product or of upgrading it. On upgrading, the product is transformed, thus giving it a new operational status which is reflected in new product functions. Specific modifications, such as alterations in the software or hardware, are carried out on the product to equip it with better, extended or new functional features compared to its original condition. Consequently, the product can be improved, extended or utilised to perform completely new tasks. When choosing upgrading, a product almost starts a new life.However, technical or economic circumstances sometimes mean that a product cannot be upgraded. To change this situation, far-sighted product planning is required which commences in the product-engineering stage. In this early phase of development, the fundamental product features - including later modification possibilities - are fixed. Improvements in utilisation performance are mainly suitable for high-quality capital products with an intensive usage phase such as heavy vehicles, machines, electric and electronic equipment (EEE), etc. For these products, service and maintenance are frequently required during the usage phase to ensure normal functionality.The appropriate acquisition and usage of product field data can provide a basis for improving the utilisation performance of:
- current products by continuously monitoring their condition and status and
- the next generation of similar products through design improvement of aspects related to usage performance such as reliability, maintainability, availability, etc.

The achievement of these goals is dependent upon various tasks such as the recording, integration, management and analysis of product field data. This provides the necessary information and knowledge which can be used as a basis for making decisions or solving problems related to predictive or corrective maintenance, design changes, production adaptation, etc.

4.4 Product data management for high data continuity

Appropriate strategies, methods and tools need to be applied in order to reduce the lack of information in the early stages of a product's life cycle. Here, qualified activities and systems must be capable of transferring know-how and information, for example with the help of communication networks based on reference models and simulation tools.

"The digital factory" ensures high data continuity for life cycle management and improves and accelerates all simultaneous engineering activities. Appropriate computer-aided tools for planning, analysis, assessment and simulation are used in order to include manufacturing data in early reuse studies, design reviews or manufacturing planning. A concurrent exchange of information expands interoperability and enables a higher degree of communication between specialists.

From the moment of the development of concept for a manufacturing system onwards, many stakeholders are involved though suppliers, purchasing, installation and use of a system. To achieve effective interaction and communication, open information through standards (Kjellberg et al. 2004) is becoming a prerequisite. The main standards of interest are STEP AP 214 (Johansson 2001) and its subset in the PDT schema. To manage manufacturing life cycle information and manufacturing requirements, it is also important to consider such standards as AP 239, PLCS, Product Life Cycle Support and AP 233 for Systems Engineering. A focus has been placed on the "Guideline for the Usage of Data Formats for Describing Production Equipment" and its use during the life cycle (Nielsen 2003). The authors of the book have contributed to the development of the standard 13399 for mall tool information management together with Sandvik and Kenna Metal (Nyqvist 2006) The relationship between the three main domains: product, processes and resources (PPR) is becoming increasingly important.

As a result of regulations, increasing functions and the complexity of modern manufacturing systems, the product stewardship of a manufacturer is expanding. To ensure maximum performance and secure processes during usage, the manufacturer becomes more and more involved at this stage. The user or operator of a machine is supported by means of tele-service and tele-operations controlled by the manufacturer. All these activities must follow the law of maximum economy. Optimising a life cycle by assessing economical criteria requires the application of new cost-accounting methods in order to determine the share of costs and revenues.The close cooperation of all business partners involved in a product's life cycle is a prerequisite for optimising the design and operation of a product. To this end, configuration management and documentation must be organised in a way which takes the needs of the different life cycle partners into account. The organisation of documentation and data is essential for a clear and unambiguous product configuration at all stages in a life cycle as well for the realisation of efficient technical support processes within life cycle management. Thus it is crucial that the same data are available for all the activities performed by the different life cycle partners in the various phases of the product life cycle.

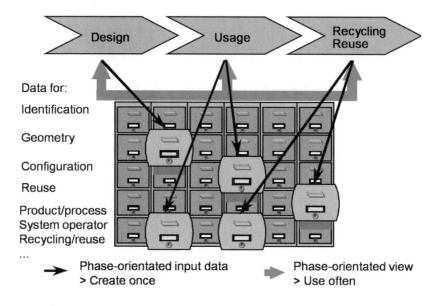

Fig. 4.13: Integrated data model.

An integrated information model is the key technical factor in determining the success of technical support processes (Figure 4.13). It is an information reservoir of the entire life cycle and consists not only of complete product data but also of data which are not directly related to the product but necessary for competent consulting in technical support processes. Reference models for the life cycle phases of the different life cycle partners, for the documents and data allocated to processes in these phases and also models of useful cooperation processes may all help to speed up agreements on life cycle management between partners and to implement an integrated information model.

Once the informational needs of the cooperation processes have been ascertained, an integrated information data model can then be conceived. For this, the detailed identification of data transferred with the documents in the cooperation processes is essential. These activities and structures result in fundamental challenges in terms of information consistency, redundancy, reliability, efficiency and security.

4.4.1 Using field data to close information loops

An important issue in PLM is the analysis of the different product life cycle phases. This is necessary in order to be able to translate and transform the related data/information into information and knowledge for varied use by different actors in various phases of the product life cycle. This section of the book addresses this aspect by presenting the main research issues regarding the closing of information loops within and between the different phases of a product's life cycle in the PROMISE project.

A product system's life cycle is characterised by three main phases: (i) beginning of life (BOL), including design and production, (ii) middle of life (MOL), including use, service and maintenance and (iii) end of life (EOL), characterised by various alternatives such as reuse of the product with refurbishing, reuse of components with disassembly and refurbishing, material reclamation without disassembly, material reclamation with disassembly and, finally, disposal with or without incineration (Kiritsis et al. 2003).

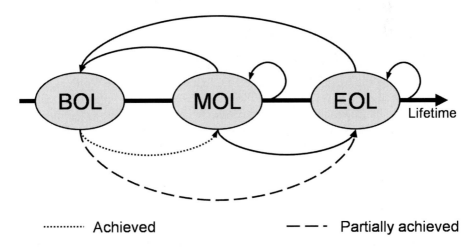

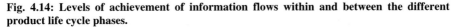

Fig. 4.14: Levels of achievement of information flows within and between the different product life cycle phases.

The overall objective of PROMISE (Product Life cycle Management and Information Tracking Using Smart Embedded Systems), an integrated project of the Sixth Framework Program, was to develop a new generation of product information tracking and flow management system. This system enables all actors playing a role during the life cycle of a product (managers, designers, service and maintenance operators, recyclers, etc.) to track, manage and control product information at any phase in its life cycle (design, manufacturing, MOL, EOL), at any time and any place in the world (Kiritsis et al. 2003). Figure 4.14 shows the information flows which are achieved, partially achieved or not achieved at all. The full arrows show the information flows which are not achieved. These are worthy of investigation and are intensively considered in the PROMISE project. The information flows between design and production in BOL are not represented because they are well-established.

In many companies, there is quite a substantial history of the acquisition and analysis of product life cycle data. The recording of product life cycle data is facilitated by the enormous progress made in the development of sensors and other product-embedded information devices (PEIDs) such as Radio Frequency Identification tags (RFIDs) and measurement technology. However, the primary purpose of these data has been to use them for managerial, marketing or logistics purposes. The transformation of such data into appropriate information and knowledge can be useful for many other reasons, such as corrective maintenance, predictive maintenance, improving design and production, determining the best EOL alternatives for treating retired products, etc. This type of use of product field data significantly contributes towards one of the main objectives of the PROMISE project, i.e. the closing of product life cycle information loops.

BOL data provide data about products and their structure. Examples of BOL data are product identification, material codes, BOM, manufacturer identification, manufacturing date, instructions for dismantling and recycling, etc. These data are static and should be made available throughout the whole product life cycle.

MOL data provides information related to the use, service and maintenance of products during the usage phase. Unlike BOL data, MOL data are highly dynamic and keep increasing over time. In the past, these data were mostly entered manually into the corresponding field database by individuals such as maintenance personnel. Nowadays, this task can be done automatically and in real time using PEIDs such as RFIDs and sensors attached to the products and their components. An example of MOL data is the history of maintenance and repair activities, including descriptions of maintenance and repair operations and related measures such maintenance/repair time and disassembly/assembly time, list of replaced parts, information about aging statistics, maintainability problems, external conditions, etc.

EOL data are essentially as static as BOL data because there is no history of data to be maintained. The most important EOL data include the quality assessment of products and their parts/components, disassembly operations and information about later use, e.g. for re-manufacturing or recycling.

Provided that appropriate tools and methods are used, the transformation of product field data gathered during different life cycle phases can provide information and knowledge for different purposes and different life cycle phases:

- BOL:

    - Design—to improve some DFX aspects such as reliability, availability, maintainability (RAM), life cycle cost (LCC), safety, environment, etc.,
    - Production—to adapt production processes,

- MOL—e.g. in predictive maintenance, corrective maintenance, etc.,
- EOL—e.g. in decommissioning and selection of relevant EOL options.

Another approach has been realised by a French consortium under the funding of the French Ministry of Economy, Finances and Industry. The concept is titled "Integration of Product Process Organisation for engineering Performance improvement" (IPPOP). The IPPOP project is concerned with the integration of product, process and organisation related dimensions of a design project and the corresponding extension of existing CAD/CAM tools. The main technological issues of the IPPOP project originate from the usage of PDM systems and from other design conditions: multi-user / multi-abilities environment, market conditions, human and material resources, management and the capitalisation of knowledge and know-how.

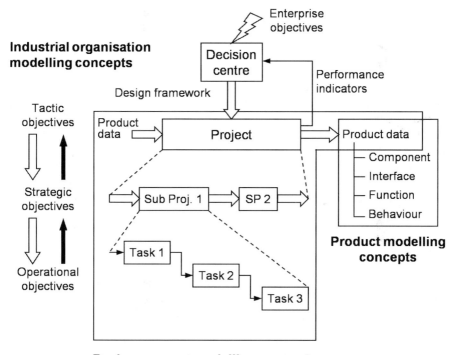

Fig. 4.15: Conceptual representation of the IPPOP model.

The IPPOP (Girard 2004) project aimed at linking the three dimensions related to generic collaborative design activity:

- Industrial organisational dimension: a company needs to be structured according to project issues and described by its human resources. Organisational modelling permits project objectives, performance and decision criteria, dedicated resources, etc. , to be defined. It manages the design process in order to maximise the fulfilment of objectives. The design process can be seen as being a product life cycle model.

- Design process dimension: this dimension formalises each design activity related to specific individual or collaborative design tasks. It is essentially based on the IDEF model supplemented by the addition of original concepts.
- Product dimension: the model represents different breakdowns and/or graphs of the product. It is based on an existing model and has been completed by adding new elements to enable multiple views, alternatives, CAD services, etc.

Figure 4.15 is a conceptual representation of this approach which has also been formalised with the aid of a UML class diagram. The diagram has been used to implement the first IPPOP demonstrator (http://ippop.laps.u-bordeaux1.fr/).

4.4.2 Enduring design records

The ability to associate product and process with one another is useful not only as a communicative and managerial tool but also as a means to enrich engineering design records. In the case of extended life cycles - in many cases beyond the careers of those involved in product creation - such records become the sole means of communicating the intent of the design to engineers involved in product support services. The KIM project, a UK EPRSC-funded research project, seeks to improve and unify such records of designed product, design process and design rationale. This makes it possible to maintain an enduring and complete record of a design in order to enable the proper support of a product. Current industrial design records generally centre around a geometric depiction of the product with little indication of *how* the product was designed (this omission is most acute at the detailed design level) or *why* it was designed in that manner (McMahon et al. 2005). As such, a company's ability to maintain, modify, upgrade or refit a product to maintain performance throughout its life is significantly impaired because the intent of the original designers and how this intent is embodied in the design can only be assumed or inferred.

At a technical level, alongside curational issues (as discussed by Ball et al. 2008), a key part of the project focuses upon providing improved product representations, accurate process descriptions mapping onto this product representation and records of the decisions and rationale which guided this process. The organisational approaches employed in making these records available to engineers unfamiliar with the original design are also considered.

Although not the sole model of a product, the CAD geometric model has become the principle mechanism for communicating the design intent to all interested parties. These CAD models are simply sets of geometry, although greater meaning may be given by the use of annotation. In such a case, the issue of view-point-dependency becomes significant as different parties have different perspectives of the product, not all of which can be represented in the same model. For example, manufacturing engineers may be interested in the machined surfaces of a component whereas structural analysts may be interested in the surfaces that are subjected to stress. Although many CAD systems support some form of annotation, none enable the level of annotation required to adequately provide for all viewpoints. The use of 'stand-off' annotation (Ding et al, 2007) allows for the annotation constructed in different viewpoints to be stored in separate files, each of which link back to the specific elements of interest in the central CAD model. It therefore becomes possible for further design records to efficiently refer to any salient aspect of the product. Where design processes make reference to these stand-off annotation files, a unified perspective of product and process may be given at a detailed level.

The capture of the design process and supporting rationale has also received attention. Where design process modelling is relatively well-supported (a number of such approaches are discussed in Chapter 3.2), these tend to be applied at a high level and constructed retrospectively. By capturing detailed activities and the information they utilise, evolve and create, it becomes possible to construct process records from the bottom up. These records may be merged by identifying where information generated within one activity served as an input to another. As design may be constructed in synchronous or asynchronous settings, it is essential that separate documentary mechanisms be used in each and effort expended in integrating these different forms of record (Giess et al, 2008a). The application of the Design Rationale Editor (DRed see Bracewell et al. 2004) allows engineers to tackle emergent issues by mapping out all potential solutions to a given issue and going step-by-step through each potential resolution to identify pros and cons. The resultant map serves as a record of the underpinning design rationale, of what was considered, what was rejected (and why) and what was ultimately adopted.

These various records need to be organised in an enduring scheme if they are to be of use over the product life cycle. This has been addressed in two forms, the first of which takes advantage of the information dependencies established during capture of the design process. This ability to trace links between records of different activities via information dependencies allows for a map of the processes to be constructed, greatly assisting an engineer in revisiting design episodes to comprehend the evolution of an aspect of the product. Although a number of techniques may be applied, topic maps (Pepper, 2002) support the automated construction of a map and (although not intrinsically so) may be used to visualise extended design episodes (Giess et al, 2008a). The second approach is based upon earlier work in faceted classification (McMahon et al. 2002) as applied to engineering documentation. Traditional enumerative (hierarchical) classifications impose significant viewpoint dependency by nature of the principles chosen for hierarchical decomposition (the principles of division) and are therefore of limited use to engineers unfamiliar with the domain or project whose documentation they organise. In contrast, a faceted scheme permits a document or record to be classified against a number of different dimensions or facets, such that an engineer may browse and retrieve information according to any combination of facets that are of interest. This reduces viewpoint dependency, an important consideration over extended life cycles. The literature in this field (that of faceted classification) is predominantly evaluative as opposed to generative. Work in the KIM project has sought to provide guidelines upon which faceted schemes may be developed for application within engineering design (Giess et al, 2008b).

4.4.3 Digital Enterprise Technology for life cycle controlling

Digital enterprise technology (DET) is defined as "the collection of systems and methods for the digital modelling of the global product development and realisation process in the context of life cycle management" (Maropoulos 2003). DET is implemented by the synthesis of technologies and systems from five main technical areas, the DET cornerstones (Figure 4.16). These cornerstones are the design of products, processes, factories and technologies for ensuring conformity of the digital with the real environment as well as with enterprise design and logistics.

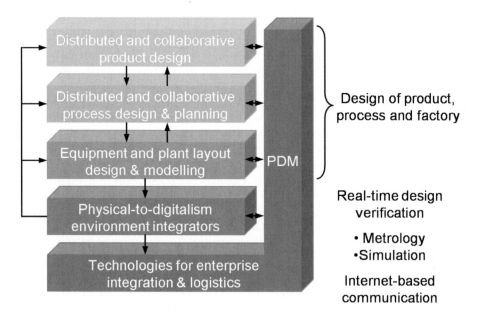

Fig. 4.16: The cornerstones of digital enterprise technology.

DET has a *heterarchical structure*, with functionality configured by the flexible, Internet-based integration of data repositories, distributed systems and user sites. The Internet is the backbone of DET, with standards for the communication and exchange of data such as STEP and XML being of primary importance. The deployment of scalable elements of DET can be configured to facilitate (a) global product development and realisation under increased (mass) customisation, (b) 'life cycle' product management and 'corporate agility' and (c) deployment of e-manufacture applications.

Life cycle controlling models will require the extension of the digital enterprise technology cornerstones into the full product life cycle. Therefore there should now be a new term: "Digital Life cycle Technology (DLT)". The new challenge is to configure software and hardware functionality to achieve full life cycle control by cutting across the barriers between all stakeholders. However, not only the software and hardware but also the basic controlling methods themselves need to be updated. In the field of cost engineering, cost estimation and cost control methods must be fully revised in order to operate in the context of the highly uncertain complete life cycle. Even within the manufacturing function, conventional cost management methods have been superseded by methods such as activity-based costing and lean accounting. Today's life cycle cost management must be used to drive product use phase services and end-of-life activities (environmental accounting) as well as manufacturing. In the new context of the complete life cycle, a significant number of functions potentially need to be performed by digital technology. These functions can be inferred by studying the corresponding literature.

A new process of architecture is required in order to configure a software and hardware system capable not only of product development life cycle control but also of full life cycle control. This form of architecture will reveal the requirements for new methods, as well as the *digital life cycle technology* components which are necessary to control the complete life cycle of products. There is potentially a very broad spectrum of functions comprising new software and hardware systems distributed across the complete life cycle of the product.

4.4.4 Examples of life cycle controlling functions

Emblemsvag uses Activity Based Costing (ABC) for complete life cycle cost management (Emblemsvag 2003). The concept of a Bill of Materials (BOM) is hence replaced by the concept of a Bill of Activities (BOA) from the use and disposal phases. The high uncertainty in cost information for the entire life cycle calls for the use of a Monte Carlo simulation approach and cost-risk distributions.

Hayek used a Weibull distribution during simulation to predict the reliability of rotary components (El Hayek et al 2005). The life cycle cost was implied from the reliability of engines and the engineering logic from their configuration. Palmer and Davis (Palmer and Davis 2005) explain the deficiencies of performance measures such as ROI (Return On Investment). For example, "the benefits of FCIM - increased production capacity, improved quality, reduced inventory, reduction in labour, improved cycle time and safety and increased flexibility – are at best difficult to quantify". The research highlights the danger of simplicity in cost engineering in this case. The benefits of investing in Flexible Computer Integrated Manufacturing (FCIM), for example, are underestimated. Investment in FCIM is poor in the USA compared to Germany and Japan.

Value Orientated Life Cycle Costing (VOLCC) was introduced by Janz and Sihn (Janz and Sihn 2005) as a concept for optimising designs for the complete life cycle. Chief methods used are: value analysis, Quality Function Deployment (QFD) and Life Cycle Costing (LCC).

Wise approached the servicing of aircrafts from a design point of view (Wise et al 2005). The concept was Maintenance Task Analysis (MTA) to enable the cost-effective design of aircrafts in order to save money during the support phase (maintenance and support are quoted as 60-75% of full life cycle costs).

Aurich and Barbian examined the life cycle cost of production systems because of "fast changing markets and ever shorter product life cycles" (Aurich and Barbian 2004). The idea of flexibility was designed using a *production project master schedule, process modules* and a tool called "Flexible Mode and Effect Analysis" (FlexMEA). The idea of life cycles can be associated both with products and production systems.

Neag examined the aspect of Commercial Off The Shelf (COTS) software which improves the life cycle cost of *automatic test system solutions* (Neag 2004). These improving aspects show to be "modular, based on correct functional allocation, should possess distribution capabilities and contain open interfaces that remain backwards compatible".

Fenton et al. reviewed the field of automated intelligent diagnosis (Fenton et al. 2001). As fault diagnosis requires advanced skills, automated tools could make reduce costs. Examples of intelligent tools include: "rule-based, model-based and case-based".

Al-Najjar and Alsyouf stated that there is a simple dichotomy in maintenance costs, namely direct and indirect costs (Al-Najjar and Alsyouf 2004). Simplicity continues in the difficulty in estimating indirect costs, e.g., "loss of income due to breakdowns, poor quality, loss of customers and market share". The research stated in its conclusions: "in a recession greater investment should be made in maintenance rather than reducing its budget, because investing in maintenance can return nine times the invested capital over the depreciation period". The research developed a data collection sheet requesting both technical and economical data were not explicitly available in company databases. Discussion of data sheet concepts and their context was considered to be relevant. The idea of maintenance performance measures and cost factors was especially important. These factors and measures included the notion of maintenance profit which was defined as: "the difference between the maintenance savings estimated during a certain period, e.g. one year and the investments made in maintenance for improving maintenance efficiency, productivity and the company's profitability during the same period".

Chiu and Chiu introduced the method of *time-based competition* in which work completion time is important for the highly-uncertain manufacturing environment

(Chiu and Chiu 2005). The important concepts are concerned with decision-making for production method and supplier selection, involving measures of price, on-time delivery, quality, flexibility and performance. A literature review by Chiu and Chiu found a supplier performance index which was used to reflect the logic of the buyer-seller relationship (Chiu and Chiu 2005). Further methods found in the literature involved the Critical Path Method (CPM), dynamic programming logic, mixed integer method and a matrix called the Design Structure Matrix (DSM).

4.5 Interim summary of life cycle information support

The receding chapter showed how the sustainable design of product life cycles can be effectively supported using continuous information support. Product usage data provide knowledge gained from experience which can be used in product development to optimise future product generations and ensure their sustainability. Permanent data analysis also enable users to monitor products continuously when in use and to generate optimisation approaches from the various data sources. To achieve this, the data must be entered into the appropriate life cycle model (see Chapter 2) and be made available for continuous life cycle evaluation (see Chapter 3). Therefore, the integration and utilisation of these sources of knowledge throughout the entire product life cycle represent an enormous potential for the sustained optimisation of products and the attainment of durable competitive advantages.

4.6 References concerning chapter 4

(Al-Najjar and Alsyouf 2004) Al-Najjar, B., Alsyouf, I.: Enhancing a company's profitability and competitiveness using integrated vibration-based maintenance: a case study, European Journal of Operational Research, 157, pp.643-657, 2004

(Aurich and Barbian 2004) Aurich, J.C., Barbian, P.: Production projects – designing and operating lifecycle-oriented and flexibility-optimized production systems as a project, International Journal of Production Research, 42(17), pp.3589-3601, 2004

(Ball et al. 2008) Ball, A., Ding, L. and Patel, M. 2008. An approach to accessing product data across system and software revisions. *Advanced Engineering Informatics*. Vol 22, Issue 2. pp 222-235

(Berger et al. 1998) Berger, R., Krüger, J., Neubert, A.: Internet-basierter Teleservice; in: Industrie Management 6/98, GITO-Verlag, Berlin, 1998

(Bracewell et al. 2004) Bracewell, R.H., Ahmed, S. and Wallace, K.M., 2004. *DRed and design folders: a way of capturing, storing and passing on - knowledge generated during design projects* in Design Automation Conference, ASME Design Engineering Technical Conferences, Salt Lake City, Utah, USA, 2004.

(Brock et al. 2001) Brock, D. L., Milne, T. P., Kang, Y. Y., and Lewis, B.: The physical markup language-core components: time and place, White paper, Auto-ID center, 2001.

(Cantamessa and Valentini 2000) Cantamessa, M.; Valentini, C.: Planning and managing manufacturing capacity when demand is subject to diffusion effects, International journal of production economics, Jul 2000; 66 (3), S. 227-240.

(Chiu and Chiu 2005) Chiu, Y.P., and Chiu, S.W.: Incorporating expedited time and cost of the end product into the product structure diagram, International Journal of Machine Tools and Manufacture, 45, pp.987-991. 2005

(Coit and Dey 1999) Coit, D. W., Dey, K. A.: Analysis of grouped data from field-failure reporting systems, Reliability Engineering and System Safety 65 (1999) 95-101.

(Ding et al. 2007) L. Ding, L., Ball, A., Matthews, J.,McMahon, C.A., Patel, M. 2007 Product Representation in Lightweight Formats for Product Lifecycle Management (PLM), *Proceedings of DET2007, 4th International Conference on Digital Enterprise Technology*, 19-21 September 2007, pp 87-95

(Dowlatshahi 2001) Dowlatshahi, S.: The role of product safety and liability in concurrent engineering, Computers and Industrial Engineering 41 (2001) 187-209.

(El Hayek et al. 2005) El Hayek, M., Van Voorthuysen, E., and Nelly, D.W.: Optimising Life Cycle Cost of complex machinery with rotable modules using simulation, Journal of Quality in Maintenance Engineering, Vol 11, No 4, pp.333-347, 2005

(Emblemsvag 2003) Emblemsvag, J.: Life cycle costing: using activity-based costing and Monte Carlo method to manage future cost and risks. Hoboken, New Jersey, USA : Wiley, 2003

(Eversheim and Schuh 1999) Eversheim, W. (Hrsg.); Schuh, Günther (Hrsg.): Gestaltung von Produktionssystemen. Berlin u.a.: Springer, 1999 (Produktion und Management 3)

(Evers and Kasties 1994) Evers, H.; Kasties, G.: Differential GPS in a real time land vehicle environment-satellite based van carrier location system, IEEE Aerospace and Electronic Systems Magazine, vol. 9, no. 8, pp. 26-32, 1994.

(Fenton et al. 2001) Fenton, W.G., McGinnity, T.M., and macGuire, L.P.: Fault diagnosis of electronic systems using intelligent techniques: a review, IEEE Transaction on Systems Man and Cybernetics, Part C- Applications and Reviews 31 (3), pp.269-281, 2001

(Foley 1999) Foley, J.: An infrastructure for electromechanical appliances on the internet, BE and ME thesis, Massachusetts Institute of Technology, May, 1999.

(Girard 2004)GIRARD Ph., "La conception: satisfaction des specifications vs conduite des activités humaines", 17ème Congrés Français de Mécanique, Troyes, Septembre 2004.

(Giess et al, 2008a) Giess, M. D., Conway, A. P., McMahon, C. A. and Ion, W.J. 2008a. The Integration of Synchronous and Asynchronous Design Activity Records. *Proceedings of 10ᵗʰ International Design Conference (Design2008)*, Dubrovnik, Croatia. May 19-22, 2008

(Giess et al, 2008a) Giess, M. D., Conway, A. P., McMahon, C. A. and Ion, W.J. 2008a. The Integration of Synchronous and Asynchronous Design Activity Records. *Proceedings of 10ᵗʰ International Design Conference (Design2008)*, Dubrovnik, Croatia. May 19-22, 2008

(Huvio et al. 2002) Huvio, E., Grönvall, J., and Främling, K.: Tracking and tracing parcels using a distributed computing approach, Proceedings of NOFORMA' 2002 conference, Trondheim, Norway, 12-14 June, 2002.

(IMITI 2002) IMTI Inc.: Modeling and Simulation for Product Life-Cycle Integration and Management," White Paper, 2002.

(ISO 1999) ISO/TC184/SC4, 1999, Industrial Automation Systems and Integration – Product Data Representation and Exchange – Part 235: Application Protocol: Materials Information for Product Design and Validation, International Organization for Standardization, ISO/NWI 10303-235.

(ISO 2005) ISO/TC184/SC4, Industrial Automation Systems and Integration – Product Data Representation and Exchange – Part 239: Application Protocol: Product Life Cycle Support, International Organization for Standardization, ISO/CD 10303-239, 2005.

(Janz and Sihn 2005) Janz, D., and Sihn, W.: Product redesign using value oriented life cycle costing", CIRP Annals Manufacturing Technology, 54(1), pp.9-12, 2005

(Johansson 2001) Johansson, M.: Information Management for Manufacturing System Development, Doctor Thesis, Computer Systems for Design and Manufacturing, KTH, 2001.

(Kiritsis et al. 2003) Kiritsis, D., Bufardi, A., Xirouchakis, P.: Research issues on product lifecycle management and information tracking using smart embedded systems, Advanced Engineering Informatics 17(2003) 189-202.

(Kjellberg et al. 2004) Kjellberg, T. Euler-Chelpin, Astrid von, Holmström, P., Larsson, M.: Guideline for the usage of data formats for describing production equipment. Computer system for Designm and Manufacturing, KTH, 2004

(McMahon et al. 2002) McMahon, C., Crossland, R., Lowe, A., Shah, T., Williams, J.S. and Culley, S., 2002 No zero match browsing of hierarchically categorized information entities, *Artificial Intelligence for Engineering Design, Analysis and Manufacturing*, Vol 16, pp 243-257

(McMahon et al. 2005) McMahon, C.A., Giess, M.D. and Culley, S.J. 2005. Information management for through life product support: the curation of digital engineering data. *International Journal of Product Lifecycle Management* Vol 1 Issue 1 pp 26-42

(Maropoulos 2003) Maropoulos, P.G.: Digital Enterprise Technology-defining perspectives and research priorities, International Journal of Computer Integrated Manufacturing, Vol. 16, No. 7/8, pp. 467-478, 2003

(Müller and Krämer 2001) Müller, P.; Krämer, K.; REFA Verband für Arbeitsgestaltung, Betriebsorganisation und Unternehmensentwicklung: BDE mDE Ident Report 2001: Das Handbuch zur Datenerfassung mit Marktübersicht, Anwenderberichten, Anbieterverzeichnis und Auswahlsoftware; Betriebsdatenerfassung, mobile Datenerfassung, Ident-Techniken (Auto-ID). Darmstadt, 2001 (Edition FB/IE)

(Neag 2004) Neag, I.A.: COTS software design minimizes ATS life cycle cost, IEEE Aerospace and Electronic Systems, 19(6), pp.29-34, 2004

(Nielsen 2003) Nielsen, J.: Information Modelling of Manufacturing Processes: Information Requirements for Process Planning in a Concurrent Engineering Environment, Doctor Thesis, Computer Systems for Design and Manufacturing, KTH, 2003.

(Niemann 2003) Niemann, J.: Ökonomische Bewertung von Produktlebensläufen- Life Cycle Controlling. In: Bullinger, Hans-Jörg (Hrsg.) u.a.: Neue Organisationsformen im Unternehmen : Ein Handbuch für das moderne Management. Berlin u.a. : Springer, 2003, S. 904-916

(Niemann et al. 2005) Niemann, J.; Österle, M.; Westkämper, E.: Erfahrungskurvenbasierte Investitionsplanung : Integration industrieller Lerneffekte in die Kostenplanung. In: wt Werkstattstechnik online 95 (2005), Nr. 7/8, S. 564-568

(Niemann et al. 2001) Niemann, J.; Galis, M., Abrudan, I., Stolz, M.: The Transparent Machine. In: Parsaei, H. R. (Hrsg.) u.a.; Integrated Technology Systems: Design and Manufacturing Automation for the 21st Century: 5th International Conference on Engineering Design and Automation, 5-8 August, 2001, Las Vegas, USA. Prospect (KY), USA, 2001, 180-185.

(Niemann and Westkämper 2005) Niemann, J.; Westkämper, E.: Dynamic Life Cycle Control of Integrated Manufacturing Systems using Planning Processes Based on Experience Curves. In: Weingärtner, Lindolfo (Chairman) u.a.; CIRP: 38th International Seminar on Manufacturing Systems / CD-ROM : Proceedings, May 16/18 - 2005, Florianopolis, Brazil. 2005, 4 p

(Nyqvist 2006) Nyqvist, O.: Information Management for Cutting Tools, information models and ontologies - Doctor Thesis in preparation, Computer Systems for Design and Manufacturing, KTH, finished spring 2006.

(Oh and Bai 2001) Oh Y.S., Bai, D.S.: Field data analyses with additional after-warranty failure data, liability Engineering and System Safety, Volume 72, Issue 1, April 2001, Pages 1-8.

(Palmer and Davis 2005) Palmer, R.J., and Davis, H.H.: Cost Accounting for rational FCIM investment", Journal of Manufacturing Technology Management, Vol 16, No 3, pp.254-264, 2005

(Parlikad et al. 2003) Parlikad, A. K., McFarlane, D., Fleisch, E., and Gross, S.: The Role of Product Identity in End-of-Life Decision Making, White paper, Auto-ID center, Institute for manufacturing, Cambridge, 2003.

(Pritschow et al. 1998) Pritschow, G. u. a.:Tendenzen in der Steuerungs- und Antriebstechnik; in: wt Werkstattstechnik 88 (1998) 1/2; Springer VDI Verlag, Düsseldorf.

(Rehmann and Guenov 1998) Rehman, S., Guenov, M. D.: A methodology for modelling manufacturing costs at conceptual design, Computer and industrial engineering, Dec 1998; 35 (3-4), S. 623-626.

(Pepper 2002) Pepper, S., 2002, The TAO of Topic Maps [Online] http://www.ontopia.net/topicmaps/materials/tao.html [accessed 1 March 2008]

(Schroer 2002) Schroer, R.: Revolutionary ideas in test, Aerospace and Electronic Systems Magazine, IEEE Volume 17, Issue 6, June 2002 Page(s):36 – 41.

(Spath 2003) Spath, D. (Hrsg.): Ganzheitlich produzieren. Innovative Organisation und Führung. Stuttgart: LOG_X Verlag GmbH, 2003

(Statistisches Bundesamt 2004) Statistisches Bundesamt: Statistisches Jahrbuch 2004 für die Bundesrepublik Deutschland. Wiesbaden, 2004

(Udoka 1991) Udoka, S. J.: Automated data capture techniques: a prerequisite for effective integrated manufacturing systems, Computers Industrial Engineering, vol. 21, No. 1-4, pp. 217-221, 1991.

(Vachtsevanos and Wang 1999) Vachtsevanos, G., Wang, P.: An intelligent approach to fault diagnosis and prognosis, The 53rd Meeting of the Society of Machinery Failure Prevention Technology, MFPT Forum, Virginia Beach, April 19-22, 1999.

(Warnecke 1993) Warnecke, H.-J.: The Fractal Company - A Revolution in Corporate Culture. Berlin u.a.: Springer, 1993

(Westkämper 2003) Westkämper, E.: Wandlungsfähige Organisation und Fertigung in dynamischen Umfeldern, in: Bullinger, H.-J., Warnecke, H. J., Westkämper, E. (Hrsg.): Neue Organisationsformen im Unternehmen - Ein Handbuch für das moderne Management, 2. neu bearb. und erw. Auflage, Berlin u.a.: Springer Verlag 2003

(Westkämper 2004) Westkämper, E.: Das Stuttgarter Unternehmensmodell: Ansatzpunkte für eine Neuorientierung des Industrial Engineering. In: REFA Landesverband Baden-Württemberg: Ratiodesign: Wertschöpfung - gestalten, planen und steuern. Bodensee-Forum. 17. und 18. Juni 2004, Friedrichshafen. Mannheim, 2004, S. 6-18

(Westkämper 2006) Westkämper, E.: Einführung in die Organisation der Produktion. Strategien der Produktion. Berlin; Heidelberg: Springer, 2006

(Westkämper and Dunker 2004) Westkämper, E.; Dunker, T. (Mitarb.); Universität <Stuttgart>/Institut für Industrielle Fertigung und Fabrikbetrieb (IFF); Fraunhofer-Institut für Produktionstechnik und Automatisierung IPA: Sonderforschungsbereich 467: Wandlungsfähige Unternehmensstrukturen. Lampertheim: Alpha Informations GmbH, 2004

(Westkämper et al. 2004) Westkämper, E. (Leitung); Drexler, K. (Red.); Moisan, A. (Red.); CIRP: Wörterbuch der Fertigungstechnik-Band 3: Produktionssysteme. 1. Aufl.. Berlin u.a.: Springer, 2004

(Wise et al. 2005) Wise, G.B., Lizzi, J.M., and Hoebel, L.J., (2005), "Annual reliability and maintainability symposium, 2005 proceedings: annual reliability and maintainability symposium, pp.61-66, 2005 (Qiu and Zhang 2003)   Qiu, R. and Zhang, Z.: Design of Enterprise Web Servers in Support of Instant Information Retrievals, IEEE International Conference on Systems, Man and Cybernetics, vol. 3, pp. 2661-2666, 2003.

(Wong et al. 2002) Wong, C. Y., McFarlane, D., Zaharudin, A. A., and Agarwal, V.: The Intelligent Product Driven Supply Chain, IEEE International Conference on Systems, Man and Cybernetics, vol. 4, 6-9 October, 2002.

(Xie et al. 2002) Xie, S.Q., Huang, H., Tu, Y.L.: A www-based information management system for rapid and integrated mould product development, International journal of advanced manufacturing technology, 2002; 20 (1), S. 50-57.

5 Customer supply networks

The concept of life cycle management can only be economically successful if the customer is offered added value through customised services. For this, customer requirements need to be accurately noted and fulfilled.

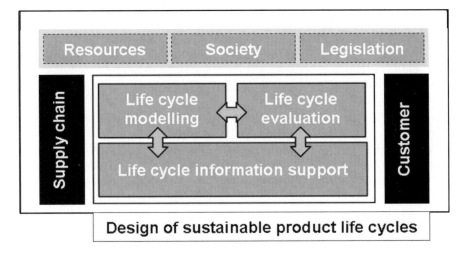

Fig. 5.1: Structure of chapter 5.

In order to master these challenges, close collaboration between all life cycle partners is required. In the future it will be essential to create efficient networks for product support in order to generate the associated life cycle benefits for customers. For this reason, the following chapter is concerned with finding ways to generate effective customer benefit through efficient supply chain management (see Figure 5.1). The optimisation of product benefit must be the focus of all network activities. For the network partners concerned, this new way of working together with the customer provides an opportunity to create additional service packages with corresponding value-adding potentials. Therefore, the chapter starts by explaining the importance of customer lifetime value. Using this as a basis, potentials for additional services are identified which can be tapped as part of supply chain networks. If the concept of customer collaboration is followed consistently, business models are created in which the product remains the responsibility of the network for its entire life cycle and only its usage (the actual function of the product) is sold to the customer.

5.1 Customer lifetime value

In an economy where the possibility of short-term access to far-reaching resources forms the basis of commercial success, the entire potential of a product's life cycle moves into the centre of strategic focus. It will no longer be a question of selling a single product to as many customers as possible, but rather of looking after a single customer and supplying him/ her with as many products as possible.

This new paradigm focuses primarily on the maximum exploitation of a single product instead of the maximisation of total sales. Businesses will concentrate more on building up long-term relationships with individual customers. Success will be measured by the amount of value-adding realised to the customer or to the product(s) sold over the total duration of the relationship (Rifkin 2000). These alliances and networked partnerships will also be given a push as a result of a dramatic acceleration in technical innovation cycles. Also, due to the fact that technical equipment, manufacturing processes, production sequences, goods and services all age faster in a highly-electronic environment, the long-term ownership of a production facility becomes less and less attractive. In order to be able to constantly maximise performance and precision, short-term access will become an ever-increasing option. Leasing, rental contracts or performance contracts will become more attractive than buying and owning. Accelerated innovation cycles and speeded-up product turnover will dictate conditions for the new network economy. Whereas the traditional market is characterised by the exchange of goods, access to holistic concepts will include material aspects in a networked economy.

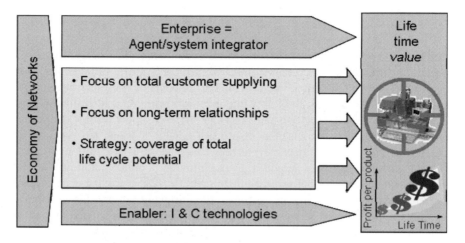

Fig. 5.2: The paradigm of lifetime value.

As a result, companies will only be able to exist if they are explicitly capable of increasing product profit by using other additional products. The paradigm of product lifetime value, i.e. the evaluation of commercial success over the entire life span of a product, therefore demands an specific focus to be placed on the requirements of individual customers (see Figure 5.2). The linking-up of products to modern information and communication technology instruments offers an excellent pre-requisite for researching and recording specific customer needs. Among other things, these technologies will also enable value-added services to be offered to the customer directly or machine optimisations to be carried out over great distances. The change in paradigm no longer places the focus of attention on the maximum use of resources implemented by companies, but rather on the maximum technical and economical exploitation of products during their life cycle. This will also be forcefully demanded due to an alteration in society's conception of value with regard to environmental compatibility and to the closed circulation of materials. The technical conditions required for this already exist and will bring massive structural changes with them. Under the pressure of international competition, it is no longer possible for many companies to survive just by manufacturing and selling goods. More and more often, enterprises are transferring added-value activities towards the areas of product design, assembly and service.

5.2 Cooperation for life cycle benefit

The customer-orientated service concept means that manufacturers are closely involved in every phase of a product's life cycle, starting with product specification. Life cycle cooperation provides the manufacturing industry with an extensive range of services which speed up and boost product development and production stages.

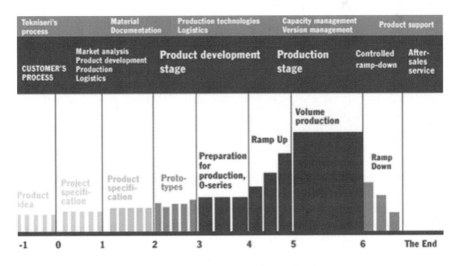

Fig. 5.3: Life cycle cooperation.

The following observations can be made from Figure 5.3:

-1-0 The customer has a new product idea or a completed product which requires special know-how from outside the company in order to finish it or complete its final exterior.

0-1 In the product specification stage, the parties agree upon details regarding different tasks, sub-areas and responsibilities which need to be taken care of as the project advances.

1-2 In the product specification stage, the customer's product is developed in cooperation with experts. At this point the product is already quite detailed and the suitability of different materials and production methods as well as different kinds of technical solutions are considered, taking cost effectiveness into account. It is here that the choices which have the most impact on a product's manufacturing costs are often made.

2-3 With the product development stage being carried out in cooperation, it is possible to reduce unnecessary testing and prototype series to a minimum, saving time and money. In order to achieve the greatest certainty of duplication and cost-effectiveness, details are polished with production.

3-4 The product development stage is the preparation for production and 0-series. During this time the product is brought towards final implementation and the design of the exterior is completed. Also, production provides additional feedback regarding processability, risk of error and lead-time. These factors can still be taken into consideration before starting serial production. Carrying out this critical stage requires close and confidential cooperation.

4-6 In volume production, production amounts are set as precisely as possible to match the customer's forecasts. Buffer stocks and storage services can be used if needed. The customer's production runs without interruption even during production peaks by using real-time and smooth logistics solutions. It is also extremely important to manage changes in product generations and the related documentation. A company's ability to invest and increase capacity is becoming more and more valuable.

6- Product support together with documentation are important for customer after-sales services. At the end of the product's life cycle, a controlled ramp down is essential. During the life cycle cooperation, valuable information is collected continuously. These data are stored and documented for future projects. This permits a successful cooperation model for future projects to be constructed as well. The data also paves the way for additional product and production optimisation services.

Surveys estimate that over the next five years, up to 70 % of the production planning and control (PPC) systems used in today's companies will be replaced by integrated order management systems. In this field alone, huge potentials exist for reducing costs by implementing new methods and techniques. In the future, a high level of customer-orientated dynamics will have to be achieved in complex net-worked systems. Long-lasting improvement will only become possible if methods as well as structures are revised and integrated into networks.

Lengthy pathways and intensified work distribution are obstacles to agility and dynamics. For this reason, the objective of market- and customer-orientated businesses must be to reduce restricting factors such as lengthy administrative procedures, non-aligned interfaces or great distances in material flows. As a result, when organising production networks, preference must be given to transformable structures with fast decision-making pathways. The virtual network process as shown in Figure 5.4 actually should be created through configuration rather than through ad hoc procedures and rules or exceedingly complex and time-consuming software design and implementation. This is because the final objective is to create and re-create efficient networks within a very short period of time. Preliminary ideas for creating such dynamics using virtual capacities already exist. However, these only permit the temporary inclusion and utilisation of resources for a short time when required. (Umeda et al. 1999).

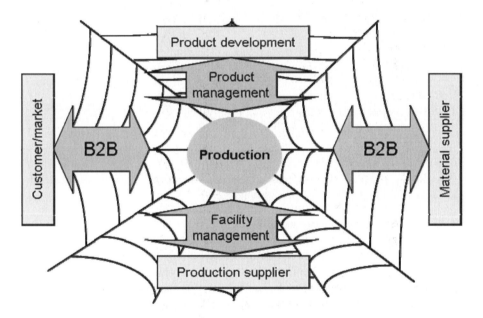

Fig. 5.4: Network partners for production excellence.

Methods for carrying out improvements exist due to the availability of faster, cheaper and, in principle, more open information and communications systems. These enable the entire flow of information from customer to supplier to be completely integrated and production is only commenced on receipt of a customer order. Additionally, formal and informal information can be made available almost anywhere in the world. The process chains run through in an order sequence could be significantly accelerated using modern means.

By networking productions, there is a huge opportunity to uphold competitiveness, as this is targeted at achieving high synergic effects and simultaneously at attaining a high level of dynamics. History has shown that successful organisations are those which keep their networks transformable and adaptable and which are able to master these networks totally. Modern information and communication technologies provide the opportunity to use these possibilities in the interest of manufacturing technology. Consequently, we should take this chance and develop it further. In the future, it will be possible to extend this system and to integrate it into a web-based platform for holistic product support concerned with all aspects of performance optimisation.

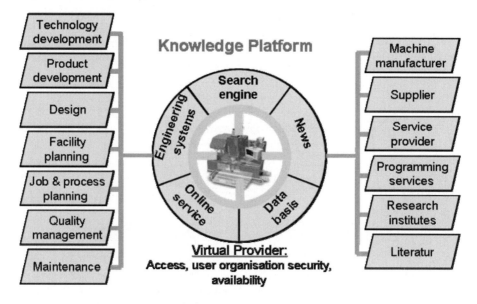

Fig. 5.5: Networks and knowledge platforms.

A web-based platform will provide the backbone for constantly-optimised machine operations (Figure 5.5). The platform will integrate the manufacturer, machine operator and various other engineering service providers right up to and including additional research institutions. These strategic alliances will accompany a product for the duration of its entire life cycle. Through this, the complete optimisation of a single product becomes the focus of attention in all business activities. The decisive criterion for future market success will be the ability to establish, organise and promote such networks for product support. These networks will be created in order to monitor a product during its entire life-time. They may be extended or reduced while any desired service can be performed on demand. The ability to control and monitor machines digitally constitutes the foundation for mastering these constraints in the digital age. On the other hand, the machine-user will benefit from such holistic networks which help to optimise his machine or machine park. The additional benefit will generate surplus profits which the user will share among the network partners.

5.3 Integrated product-service systems

Many manufacturing industries are shifting their production process from conventional mass production towards knowledge-based and service-orientated manufacturing which enables production on demand, mass customised solutions, rapid reactions to market changes and quick time-to-market for innovative solutions. It is now becoming increasingly crucial to base product development on a quick and reliable innovation process in order to overcome both the strategic and production aspects of product life cycle management. These changes on the suppliers' side are being forced due to increasingly complex and multifaceted requirements expressed by a wider group of customers (both business and consumers). The improvement demanded on the suppliers' side affects almost all levels and functions of the company, from the definition of an innovative vision/mission and the identification of a proper strategy right up the determination of consistent operative goals.

The development of integrated Product-Service Systems (hereinafter: PSS) can be seen as a new paradigm, an innovative approach as a solution to the mutations monitored in the market. A PSS can be defined as being the result of an innovation strategy, shifting the business focus to designing and selling tangible products and intangible services combined, enabling them to jointly fulfil customer needs. According to this vision, physical products (cars, white goods, mobile phones, ...) are combined with complementary services which integrate the functions and performances of the physical products (e.g.: financial plans for buying the solution, maintenance, embedded services, ...). PSSs are one of the most promising opportunities which innovative companies can rely on in order to improve their interaction with the market. They would even permit improvements in the production of standard products, which are no longer able to autonomously fulfil complex and variable customer requirements. Moreover, companies need to gain or maintain position in the market by improving not only the product itself but also the services related to the product. Finally, customers are demanding integrated solutions where products are sustained in all phases of their life cycle and where their performance standards are always at the top.

5.3.1 Developing product service systems

Especially SMES (but also many big companies) are usually focused on a specific area, while the development of integrated PSS requires a broad spectrum of competencies, skills and knowledge. This is why companies interested in the development of this kind of solution are asked either to improve their internal competences and production capabilities (hiring new employees, buying new production machinery, ...) or to improve contacts with external entities (research centres, universities, consultants or other companies) which provide complementary experiences and develop corresponding services or products. For these reasons, improvements in communication between different functions within the same company or between actors from different bodies are one of the goals to achieve. Companies are also asked to adapt their approach and internal organisation to the development of solutions in order to simultaneously manage both the tangible and the intangible side of their offer. In such a context, companies need to receive support both to add value to conventional products and also to implement completely new solutions. Therefore, the capability of already having a complete view of the product life cycle in the design phase is fundamental for the producer to enable the use of multiple Design for X strategies. Design for assembly, design for maintenance and design for environment emerged to encompass a wide range of approaches to product design. Business planning and aligning the development of new solutions with the business focus is the first stage towards successful innovation. Process planning to establish control methods and the allocation of resources is also a preliminary step towards real innovation.

The main changes which conventional companies are expected to manage in order to take up the shift towards integrated PSS include:

- Redefinition/adaptation of the innovation process in order to strengthen interactions between the different departments and actors of a company
- Redefinition of the innovation process in order to involve external entities right from the early phases in the development process
- Identification of a consistent and worthwhile strategy for facing the market
- Improvements in the organisation of the company functions
- Enhancement of activities devoted to the identification of new market opportunities and to the redefinition of company business models

Figure 5.6 shows a practical example from ABB Automation. The example clearly demonstrates how the performance of a system develops if no appropriate measures are taken to maintain efficiency over the life cycle. More and more malfunctions occur due to ageing processes, resulting in increased downtimes and greater performance losses. In order to counteract this, the company has developed a wide range of services for its customers - so-called *life cycle services* - which improve the creation of value over the entire life cycle.

The first group of traditional services reduces interruptions in production in the event of a malfunction. On the next level, preventive services help to avoid production downtimes through maintenance and regular plant overhauls keep the creation of value constant. On the top level, pro-active services increase the value of the plant through continuous modernization (e.g. upgrading and replacement).

For the customer, performance losses mean a loss in turnover which is reflected in so-called *opportunity costs*. Opportunity costs represent a measure of the amount of lost profit and describe the difference between the actual performance level of a manufacturing system and the performance level which is theoretically available on the market. Although these costs are only of a theoretical nature, they represent a significant threat if one considers the fact that potential competitors could invest in such technologies themselves. For this reason, efficient life cycle management not only provides financial advantages but also protects a company from being overtaken by competitors from a technological point of view.

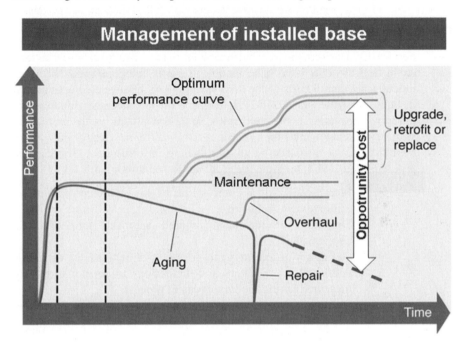

Fig.5.6: Strategies to leverage operational excellence for our customers.

If the example is considered abstractly, three fundamental strategies can be derived for developing and providing successful services or even complete service packages. The first strategy includes services which are orientated towards the entire life cycle of a product and aim at extending product lifetime (e.g. maintenance contracts). A second strategy comprises services which improve the efficiency of a product and thus increase the production yield per unit of time (performance services). The third strategy demands both dimensions to be taken into account as well as continuous services to be provided with the aim of improving performance. Only in this way is it possible to maximise product efficiency and product benefit and achieve operational manufacturing excellence.

5.3.2 Supporting activities and modules

In order to support companies in managing the improvements mentioned above, a platform offering tools and methods to innovative firms has been developed within the EC-funded RTD project ProSecCo (Product and Service Co-design, G1RD-CT-2002-00716). This platform is based on a high-level frame integrating both methodological and commercial viewpoints where ICT and non-ICT tools and methods are conceived to support SMEs in innovation management. Three main modules have been created in order to provide full assistance to companies interested in developing integrated solutions:

– A *diagnosis* module is meant to support the whole PSCD (Product and Service Co-Design) process by assessing needs in innovation fields and evaluating the PSS ideas. The module is partly integrated into the other modules and major parts of the assessment questionnaires are implemented in and accessible via the ICT platform. With the diagnosis module, the platform provides both companies and consultants with a logical, structured framework for analysing the different functional areas, identifying real problems and establishing priorities for solving the problems detected. Four main areas are addressed through this set of tools:

  • Identification of present status of the company and major improvements required
  • Definition of the most tailored approach for improving performance
  • Definition of an intervention plan, with a fixed list of results and a set of indicators and dimensions to be monitored and measured during the improvement process
  • Current and retrospective evaluation of achieved objectives and analysis of possible derivations from planned times, costs and/or quality

– An opportunity recognition module to assist firms in finding innovative PSS ideas. This is particularly important from a strategic and development point of view and is dedicated to extended market analysis, future scenario definition and idea generation (see Chapter 6.5 and 6.6). It is specifically targeted for PSS according to the results of the diagnosis and applies defined methodologies for the creation of PSS. Two main sub-modules form the opportunity recognition module:

  • The *mind setting* sub-module explains a certain way of thinking and other conditions needed to set the scene for the workflow and the process of defining opportunities. In order to be able to go through the process, a different way of thinking is required which enables a different perspective to be viewed.
  • The *methodology and process* sub-module explains the proposed steps to be taken for the process of opportunity detection. The steps provide a path to organise the thinking process in order to achieve innovative insights for product-service solutions

– A process implementation module supports the user from an organisational point of view during the implementation procedure in order to re-define the approach of the company to the innovation process and organisation according to the requirements of the new PSS. The main activities performed by this module include:

  • Acquisition and analysis of current practices
  • Formalisation of present innovation process
  • Proposal of improvements through the involvement of new actors
  • Definition of a new innovation process
  • Testing and simulation

The results achieved in the project enable new PSSs to be generated where companies intend to innovate beyond the state of art. It also targets existing systems if improvements both of the system and the organisation are required. The experiences gained with SMEs and large companies confirm that PSSs can be successfully integrated if tangible and intangible assets are properly structured and managed by the company. Their co-design patterns are strictly linked to organisational processes. The need to reformulate already-existing working schemes through successful re-engineering often reflects new organisational and technological structures, the management of change and motivation, the need for training and the commitment of people to an open vision regarding the entire life cycle of the product. Companies are rarely interested in adopting the whole range of solutions simultaneously: step-by-step changes and improvements are usually favoured. Companies with a valid product and solution portfolio generally prefer to improve their innovation process in order to better exploit their technology and solutions, especially through the involvement of customers and other partners in their innovation process. SMEs are the main customers: they usually lack management and innovation management skills and this tool supports them in finding the right complementary partners. Large companies on the other hand tend to use only some of the tools, coupling them with approaches, methodologies and systems they have already adopted.

5.4 Selling the benefit instead of the equipment

The fields involved with the initial steps of processing basic materials or manufacturing parts, components and equipment are being dislocated. By using information and communications systems, suppliers become involved in product development. However, the so-called *system management* will remain in the hands of the OEMs operating directly with the market. The field of after-sales where long-term customer relationships can be established is gaining strategic importance.

This development can be extended right up to the level of so-called performance contracts. Here, the manufacturers of technical products will also take over operation of their products and will only sell their usage (Figure 5.7).

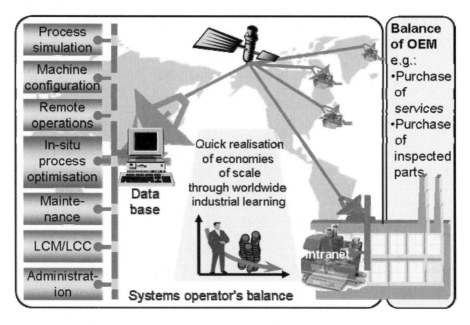

Fig. 5.7: Systems operators and performance contracts.

(Westkämper et al. 2001) This not only leads to increased manufacturer responsibilities but also to a stronger and more durable relationship with the OEM. The OEM, for example, purchases the service (instead of the machine) or pays only for inspected parts. This implies a conversion of fixed costs into variable costs for the OEM along with the extension of the value-adding chain and an opportunity for the manufacturer to increase profits. The manufacturer becomes a system operator who offers his services worldwide. All machines are connected and operated by modern information and communication technologies and controlled in a central surveillance centre where all incoming data are acquired, analysed and evaluated.

The availability of these tools will enable system operators to benefit from economies of scale. Once a database has been set up, statistical performance evaluation becomes possible and the *best practice* can be determined. Through world-wide learning, system operators will be able to rapidly ascend the learning curve. By applying information and communication technologies, new knowledge will become immediately available which can be implemented on all machines worldwide. (Niemann 2003a), (Niemann 2003b), (Niemann 2007) (Niemann et al. 2004, (Niemann and Westkämper 2005)

5.5 Industrial prototypes and practical examples

5.5.1 Example of the implementation of LCC methodology

This section is concerned with an application of LCC within the EC-funded IST project PROMISE (Product Life cycle Management and Information Tracking using Smart Embedded Systems - 507100). The methodology has been applied here to evaluate advantages deriving from the introduction of a new predictive maintenance approach for machine tools. Implementation of the LCC methodology requires customisation and adaptation to each specific product/situation in order to make the model work efficiently. The most important common phases include:

- Definition of the problem: each situation, industry, company and product requires individual examination and a *customised* LCC model
- Identification of feasible alternatives: in order to quicken the analysis and response of the LCC tool, only the most promising alternatives are taken into account
- Development of Cost Breakdown Structure (CBS): for each alternative a (top-down) hierarchical structure spreading budgeted resources into the elements of costs (labour, materials and other direct costs) is defined.
- Selection of a cost model for analysis: the *net saving* is the computed indicator in the chosen approach to calculate the difference between the LCC of each alternative and the LCC of a pre-defined standard-option.

$$NS = LCC_{option} - LCC_{STD}$$

- Development of cost estimates and cost profiles: during this step, a rigorous analysis of the different costs and their estimation is performed. In order to produce a reliable set of data, a Monte Carlo-based simulation can be performed and data and errors gathered and stored. In particular, the model considers an estimation of the (cash flows) both for the company and for the (potential) user.

The example below shows cash flow variations for a machine-tool manufacturing company which adopts an innovative maintenance approach. In the first row, Δ(cash flows) for the machine manufacturer are shown, while in the second row, Δ(cash flows) for a machine-tool purchaser are calculated (for a generic year):

$$\Delta CF_{suppl} = I + \Delta C_{manpower} + \Delta C_{production} + C_{failed\_production} + \Delta Sellings_{broken} + \Delta Sellings_{setup}$$
$$+ \Delta C_{dismissal}$$

$$\Delta CF_{purchaser} = \Delta price + \Delta C_{installationr} + \Delta C_{manpower} + C_{training} + \Delta C_{rejects} + C_{failed\_production} +$$
$$+ \Delta Sellings_{broken} + \Delta Sellings_{setup} + \Delta C_{Ter\min alValue}$$

- These formulas are applied for each year of the selected time-horizon and for each alternative.
- Break-even analysis: break-even and trade-off charts are drawn to support decisions and choices between different alternatives.
- Identification of high cost contributors: this analysis only permits quick action to be taken on a selected set of areas/activities.
- Sensitivity analysis: all the acquired data are based on estimations. These estimations may originate from analogies with other similar (past) products, parametric models of the reality or from industrially-engineered procedures (standard costs for each alternative). In all cases, data are composed of an *absolute value* and of an error affecting this value. On performing a sensitivity analysis, the decision-maker would be able to judge the effect on the final performance derived from a variation of a specific addendum in one of the afore-mentioned formulas.
- Risk analysis to estimate the overall risk of each option.
- Recommendation of a preferred approach: the decision-maker has the opportunity to handle consistent and reliable data to express his final judgement

The introduction of a new maintenance policy has an impact on many different aspects:

- Improvement in machine performance.
- Improvement in machine tool efficiency: this does not necessary result in a longer life cycle but does enable the machine tools to maintain their maximum productive potential for a longer period
- Increase in productivity thanks to a higher working speed at the same level of quality
- Decrease in breakdowns combined with resulting savings of time and money, e.g. the realisation of a mould would demand considerable working time
- Optimal maintenance plan and a reduction in emergency interventions
- Reduction in production costs associated with losses
- Decrease in rejects and re-working
- Reduction in setting/calibration times
- Decrease in maintenance contract costs
- Test automation

In a LCC approach, all the considered issues are transferred to the economical side through the identification of cost savings and possible revenues (identifiable as negative costs) both from the producer and the machine tool user.

As previously explained, costs from the manufacturer's side can generally be grouped into the following categories:

- Research and development, production, maintenance, revenues, purchase

- Purchase, installation, manpower, training, waste, loss of production, breakdowns

The identification of all the costs belonging to each category is strictly linked to the definition of all assumptions and hypotheses affecting the scenario under consideration.

For literature references, see Dunk (Dunk 2003), Durairaj (Durairaj et al. 2002) and Woodward (Woodward 1997). Another industrial case study of applying LCA to complex industrial products is mentioned in the study of Chryssolouris et al. (Chryssolouris et al. 2001). The study analyses the environmental effects of such complex products as a commercial refrigerator. Life cycle assessment is used for analysing and evaluating the environmental performance of this complex product throughout its entire life cycle. Based on the results, design alterations and different manufacturing methods are suggested and the environmental impact of the improved design is estimated.

5.5.2 Example of online process monitoring

In order to cost control a manufacturing system, various data from different sources are required. It is clear that the mastery of a system's behaviour demands machine and machining data. These data can be easily acquired from the machine control system. This offers the opportunity to access machines remotely for data logging via the Internet or telephone lines. The relevant machine data can be extracted from the data flow and serve as one input for in-situ cost monitoring and forecast. Various research projects have shown that optimal logistics play an important role in preventing performance losses. A second input is data from parts logistics; for a controlling system to be able to take these facts into account, such data need to be integrated into the supervision system. A third group of data is directly related to the machine's environment. The order size, required quality, number of workers, calculated lead times etc. can all be taken directly from the work scheduling, bill of materials or order management. These data are static and can be extracted from various internal sources. Figure 5.8 and Figure 5.9 describe the structure of a controlling system implemented on a precision machining centre.

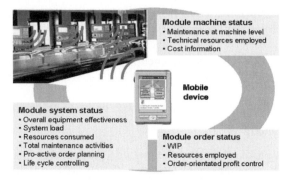

Fig. 5.8: Elements of a controlling system for manufacturing systems.

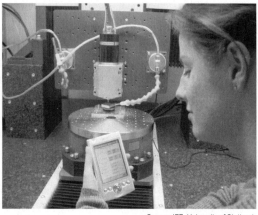

Source: IFF, University of Stuttgart

Fig. 5.9: Management and control of a micro milling facility using a mobile device.

The data are monitored and visualised using a mobile handheld PC (PDA). The mobile PDA serves as a platform for production personnel in terms of technical machine control (failures, breakdowns etc.) and economic manufacturing surveillance (e.g. deviation from estimated cost, total cost and profit, etc.). The measurement data obtained from the monitored system give a report of the actual machine status. Multiplied with cost coefficients according to the processes performed, profit analyses can be made. A sensitivity analysis of different cost positions and a comparison between machine operation times and various breakdown times serve to identify hidden performance potentials. Even a forecasting module can be integrated to simulate future profits and performance under *status quo* conditions. All data concerned with the observed machining centres must be accumulated on a top level of production programme planning in order to derive key actions in mid-term performance and resources planning.

5.5.3 Example of process monitoring for intelligent services

Two problem areas define the requirements for this service: the information required and how the information is displayed to the user. In the case of micro milling, the required process data describe the path of the milling tool.

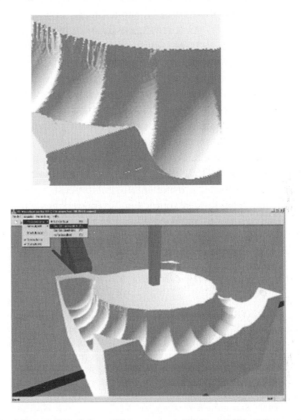

Fig. 5.10: Virtual reality model of the milling process. (Stolz and Westkämper 2005)

Therefore, the desired and actual values of the manufacturing processes can be obtained from the machine's control system (position, velocity and acceleration). In compliance with user requirements, the results of the service should be displayed to the user in a simple and ergonomic way. Such complex data need to be visualised in a manner which represents the machining process realistically. In order to achieve this, the path data are displayed in a 3-D virtual reality model which calculates the interferences between tool and workpiece in real-time. Further features such as assigning colours to certain points of the milling track give a more accurate representation. Figure 5.10 shows a screenshot of the virtual reality visualisation.

5.6 Interim summary of life cycle customer supply

The development of modern products is being decisively influenced by the application of technologies contributing towards increased efficiency. Products are becoming complex highly-integrated systems with internal technical intelligence enabling the user to utilise them reliably, economically and successfully even in the fringe ranges of technology. As a result, business strategies are aiming more and more towards perfecting technical systems, optimising product utilisation and maximising added value over the entire lifetime of a product. In this context, the total management of product life cycles associated with the integration of information and communication systems is becoming a key factor of success for industrial companies (see Chapter 4). Today's manufacturers are faced with the situation that they are required to guarantee process reliability by contract. (see Chapter 3) This implies that machines have to work efficiently over long time scales to perform different work tasks. To meet contract liabilities, machine manufacturers monitor their facilities, collect all manufacturing information and try to forecast and boost machine performance by using intelligent process optimisation. Life cycle orientated platforms provide data sources and experiences from other machines to supply information in order to attain excellence in manufacturing.

5.7 References concerning chapter 5

(Chryssolouris et al. 2001) Chryssolouris, G., K. Tsirbas, V. Karabatsou. G. Maravelakis and S. Sillis, 2001, "Life Cycle Assessment of Complex Products: An Industrial case study", Proceedings of the 34th International CIRP Seminar on Manufacturing Systems, Athens, Greece, pp. 399-406.

(Dunk 2004) Dunk, A.S.: Product life cycle cost analysis: the impact of customer profiling, competitive advantage, and quality of IS information, Management Accounting Research 15 (2004) pp. 401-414.

(Durairaj et al. 2002) Durairaj, S.K. Ong, A. Y. C. Nee, R.B.H. Tan: Evaluation of Life Cycle Cost Analysis Methodologies, Corporate Environmental Strategy, Vol. 9, No. 1 , 2002

(Niemann 2003a) Niemann, J. (2003): Ökonomische Bewertung von Produktlebensläufen-Life Cycle Controlling. In: Bullinger, Hans-Jörg (Hrsg.) u.a.: Neue Organisationsformen im Unternehmen : Ein Handbuch für das moderne Management. Berlin u.a. : Springer, p. 904-916

(Niemann 2003b) Niemann, J.: Life Cycle Management, In: Neue Organisationsformen im Unternehmen - Ein Handbuch für das moderne Management, Bullinger, H.-J., Warnecke, H. J., Westkämper E. (Ed.), 2. Auflage, Springer Verlag, Berlin u. a.; 2003

(Niemann 2007) Niemann, J. 2007. Eine Methodik zum dynamischen Life Cycle Controlling von Produktionssystemen. Stuttgart, Germany: University of Stuttgart (Dissertation). Heimsheim, Germany: Jost-Jetter.

(Niemann et al. 2004) Niemann, Jörg; Stierle, Thomas; Westkämper, Engelbert: Kooperative Fertigungsstrukturen im Umfeld des Werkzeugmaschinenbaus : Ergebnisse einer empirischen Studie. In: Wt Werkstattstechnik 94 (2004), Nr. 10, S. 537-543

(Niemann and Westkämper 2005) Niemann, J., Westkämper, E. (2005) : Dynamic Life Cycle Control of Integrated Manufacturing Systems using Planning Processes Based on Experience Curves. In: Weingärtner, Lindolfo (Chairman) u.a., CIRP: 38th International Seminar on Manufacturing Systems / CD-ROM: Proceedings, May 16/18 - 2005, Florianopolis, Brazil. p. 4

(Rifkin 2000) Rifkin, J.: Access, das Verschwinden des Eigentums, 2. Auflage, Campus Verlag, Frankfurt/New York, 2000.

(Umeda et al. 2000) Umeda, Y., Nonomura, A. and Tomiyama, T.: A Study on Life-Cycle Design for the Post Mass Production Paradigm, Artificial Intelligence for Engineering Design, Analysis and Manufacturing, Vol.14, No. 2, Cambridge University Press, pp. 149-161., 2000

(Westkämper et al. 2001) Westkämper, E., Niemann, J., Stolz, M.: Advanced Life Cycle Management in Digital and Virtual Structures. In: Chryssolouris, George (Hrsg.); University of Patras/Dept. of Mechanical Engineering and Aeronautics / Lab. for Manufacturing Systems and Automation: Technology and Challenges for the 21st Century: CIRP 34th International Seminar on Manufacturing Systems, 16-18 May, 2001, Athens, Greece. Athens, Greece, 2001, S. 1-5.

(Woodward 1997) D. Woodward – Life cycle costing theory, information acquisition and application, International Journal of project management Vol. 15 N°6 pp. 335-344, 1997

6 Method for the design of life cycle concepts

In the preceding chapters, the fundamentals of the design of sustainable life cycles were presented. The following chapters are concerned with combining the various modules to form a holistic method (see Figure 6.1).

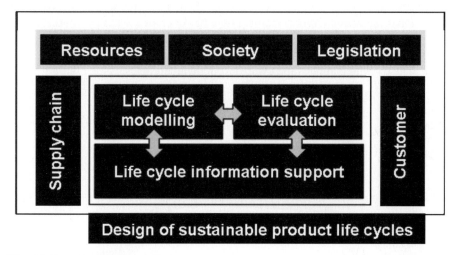

Fig. 6.1: Structure of chapter 6.

In the process, the innovative method takes into account the various stakeholders involved in the life cycle of a product. It places a particular emphasis on the integration of all life cycle partners throughout all the different phases of a product's life. Compared with methods used up till now, a life cycle orientated planning approach involves more effort in order to integrate the extended planning horizon and further use of the product on completion of its first usage phase. Therefore, the method described in the following focuses especially on the phase of product design.

6.1 Approaches from a retrospective point of view

The objective of life cycle design is to design sustainable product life cycles which minimise the consumption of energy and materials as well as the amount of waste and environmental emissions generated during the entire life cycle, while maintaining welfare and corporate profits (Umeda et al. 2000). The design of a product must be viewed as the design of the life cycle. For example, the design of a product in the earlier phases of the life cycle determines whether services supplied during the latter phases of a product's life cycle (e.g., maintenance and functional upgrading services) become either cost factors or profit factors.

Manufacturing industries traditionally provide services from marketing through sales and the aspects of reclamation, recycling and disposal of products are generally outside their scope of business. However, growing global environmental consciousness are increasing constraints on the manufacturing industry. For example, environmental legislation such as the EU directive of WEEE forces the manufacturing industry to take back and recycle their products. This demonstrates the fact that the manufacturing industry now has to consider services in these later stages of the product life cycle as new areas of value adding and market creation. Under such circumstances, the manufacturing industry could be redefined as a life cycle industry which applies life cycle design methodologies.

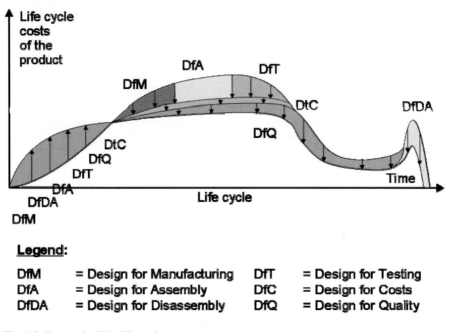

Fig. 6.2: Design for X in life cycle management.

Various concepts and Design for X (DfX, see Figure 6.2) methodologies have been proposed to support life cycle design. Examples include industrial ecology (Graedel and Allenby 1995), life cycle design (Ishii 1995), (Hata et al. 1997), life cycle engineering (Feldmann 1994), life cycle costing, design for disassembly (Boythroyd and Alting 1992), (Jovane et al. 1993), design for recyclability (Krause and Scheller 1994),(Lee et al. 1997), design for serviceability (Gershenson and Ishii 1991) and end-of-life design (Ishii 1999). Although these methodologies support certain aspects of design objects (e.g. recyclability), they provide dedicated design solutions for vertically isolated domains but do not support the holistic aspect of product life cycles. For example, the objective of recycling (especially material and energy recycling) is to reduce waste and the consumption of natural resources. Design for recycling and design for disassembly do not always contribute towards this aim, although a high degree of recyclability of a product may be obtained. Therefore, the materials used in a product life cycle should be balanced with the demand of such recycled materials. In the worst case, excessive amounts of low quality, expensive and unusable recycled materials could be created as a result of encouraging design for recycling. (Umeda 1999)

Life cycle Assessment (LCA) (Wetzel and Hauschild 1994) is a powerful tool for evaluating material and energy consumption as well as life cycle emissions from a holistic perspective. However, it cannot evaluate the balance of flows of material, energy and money, especially in a life cycle which contains closed flows due to remanufacturing, reuse and recycling processes. In order to design a sustainable product life cycle considering *design of life cycle*, designers need to model, evaluate and design flows of material, energy and money in a product life cycle while at the same time considering the balance of these flows among the life cycle processes.

To design a sustainable life cycle which may include the multiple reuse of components, the whole balance of the product life cycle has to be evaluated from a technical and economic viewpoint during the early stage of design. Although accurate information about the whole life cycle is not always available in this phase, the balance should still be evaluated. In some cases, CAD systems are able to supply precise design information about the product and PDM systems can give detailed bills of products, materials and processes which enable an accurate evaluation. However, it is rare that all such information is available at an early stage of design. Furthermore, additional kinds of information, especially about downstream processes which conventional manufacturing industries do not consider, also needs to be taken into account in order to evaluate the complete life cycle. Examples of such information include practical lifetimes of components, customer behaviour, reuse rates, collection rates and recycling rates. Therefore, it should be possible to calculate the evaluation both with and without the availability of precise data.

Information networks between the stakeholders such as suppliers, manufacturers, dealers, customers and other service providers in the entire life cycle of products enable us to acquire relevant life cycle information and to design and improve product life cycles using life cycle design tools such as the life cycle simulator. (Shu et al. 1996), (Tomiyama 1997), (Umeda et al. 2000), (Umeda et al 2000)

In general, because life cycle activities can be implemented in various ways, a product life cycle can have various options. Therefore, a decision needs to be made about selecting the best life cycle options in order to optimise product life cycle costs at an early stage of product development. In other words, the product flow during the product life cycle needs to be determined. This is known as the product life cycle strategy.

Some research work has been carried out which deals with the subject of product life cycle strategies. A pioneer in this field is Kimura. Kimura proposed the life cycle design for inverse manufacturing (Kimura 1999). He simulated life cycle performance with various environmental factors to select the best type of life cycle for a specific product design. He described four typical life cycle strategies: very rapid take-back with complete reuse, rapid take-back with upgrading in a factory, long life with upgrading performed by the customer and unlimited life with complete maintenance. Recently, Kimura and Hata (Kimura and Hata 2003) dealt with the design and management of life cycles based on a simulation of the total product life cycle.

Others have also focused on this research issue. For example, Faux et al. introduced a basic product design strategy by classifying product life cycle scenarios (Faux et al. 1998). The authors proposed a simulation of the product life cycle to evaluate the effectiveness of possible life cycles in product conceptual design. Rose et al. (Rose et al. 2000) proposed Stanford's ELDA (EOL Design Advisor) which focuses on the designer perspective and seeks to recommend a best-practice EOL strategy. Umeda and the LCDC Life Cycle Design Committee addressed the life cycle design guideline for inverse manufacturing with a focus on determining the life cycle strategy (Umeda and LCDC Life Cycle Design Committee 2001). In the process, they clarified the applicability of several life cycle options such as maintenance, upgrading, reuse and material recovery. Yu et al. described the life cycle selection problem as being a multi-objective optimisation problem and solved it using a goal programming technique (Yu et al. 2001). Kato et al. described decision factors for determining product life cycle strategies (Kato et al. 2001). They used simulation to try to find appropriate combinations of product characteristics and life cycle strategies. In addition to this, Mangun and Thurston (Mangun and Thurston 2002) presented a model for making design decisions about component reuse, remanufacturing, recycling or disposal over several product life cycles for a portfolio of products. They considered cost, reliability and environmental impacts simultaneously in order to determine when each component should be reused, remanufactured, recycled, or disposed of.

It is essential to develop a method which assists in generating, evaluating and selecting an appropriate life cycle design because this affects the performance of the overall product life cycle. Up till now, it has not been easy to implement because it requires not only the availability of product life cycle information but also knowledge of a wide spectrum of engineering disciplines (Sánchez 1998). However, thanks to emerging product identification technologies (Kiritsis and Rolstadås 2005), it is becoming a challenging issue. Consequently, a solution framework needs to be developed which determines the best life cycle strategy in order to optimise the performance of the entire life cycle while considering the requirements of the product life cycle. This solution framework is supplied by the method for the design of life cycle concepts which is described in the following chapters.

6.2 New requirements call for a new approach

As explained above, there are many methods for optimising product life cycles but few attempts have been made to construct a model for designing life cycles. Most existing methods and tools assume that there is an existing product whose life cycle can be improved with regard to profit (e.g. life cycle costing) or environmental friendliness (e.g. LCA).

A more holistic approach is product life cycle management, which has two objectives. First of all, the flow of information throughout the life cycle is managed. The second objective is to design the product in a way which makes several life cycles for the product possible, for instance through upgrading (Pahl et al. 2007). However, the development of a new life cycle concept is not supported by product life cycle management.

The development of a new life cycle concept has to be integrated into the existing product development process (Figure 6.3). Trade-offs between the different phases of the life cycle have to be taken into account already in the conceptual design (Niemann 2007). Furthermore, a new life cycle concept needs to be free of as many restrictions as possible, which is why it is assigned to the step of product planning:

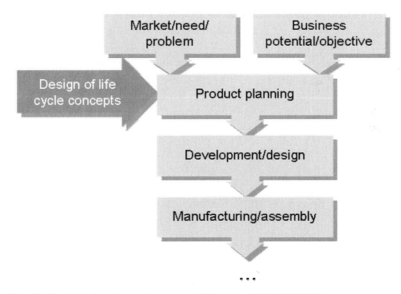

Fig. 6.3: Product development process. (following VDI 2221 1993)

In classical product planning, generally little attention is paid to the issues such as product recovery or the design of services for manufactured goods (see Pahl et al. 2007; Holzbaur 2007).

A "method for life cycle-orientated product design" was published by Kölscheid (Kölscheid 1999). As the name indicates, product design is central to his work and the implementation of design solutions his priority.

An attempt to combine life cycle management and conventional product planning methods has been made by Mateika (Mateika 2005). However, due to the detailed analysis of the present cost structure and focus on existing products, only incremental changes are likely as an outcome of this method. Also, the scope of Mateika's work was limited to the machine and plant construction industry and no regard was given to other manufacturers of technical goods (Mateika 2005).

Most conventional product planning methods propose a similar procedure (Pahl et al. 2007) for developing new product concepts. Pahl et al. (Pahl et al. 2007) list seven steps of product planning, but only the first five steps are of interest for the issue discussed here:

1. Situation analysis
2. Definition of a strategy for search
3. Finding product ideas
4. Selection of product ideas
5. Definition of products
6. Implementation planning
7. Clarification and specification

The first five basic steps are included in a certain form in the method for the design of life cycle concepts, as described in the next paragraph (see also Figure 6.4).

6.3 Methodological approach

To develop life cycle concepts, new ideas have to be found. The strategy for the search for new ideas and the fields on which the search is based are defined in Chapter 6.4 and Chapter 6.5. While Chapter 6.4 offers a classification of the measures which can be used to close open life cycle loops and optimise life cycles, Chapter 6.5 focuses on generating descriptions of the possible future state of things (Figure 6.4).

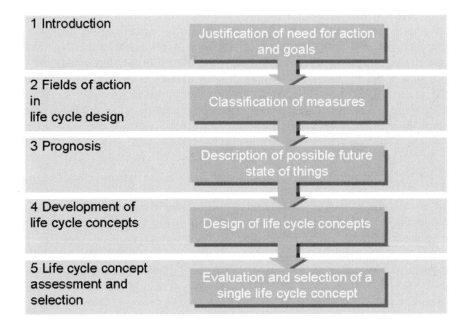

Fig. 6.4: Overview.

This classification and "pictures of the future" are used to find new ideas in the fields of action. The most promising product ideas are determined in a preliminary selection and two to three coherent life cycle concepts are then developed from these ideas (design of life cycle concepts).

The final step is for the top management to select a single life cycle concept (life cycle concept assessment and selection). To support the decision-making process, the life cycle concept is evaluated according to three criteria: costs and revenues, environmental impact and benefit for the users.

6.3.1 Objectives for the design of life cycles

Two fundamental motives to take action in order to improve product life cycles can be found in literature: ecological and economic motivations. In accordance with these, the methods for assessing product life cycles can be assigned to either one or the other of these approaches. The value, importance and reasoning of both motivations are outlined in the following section.

6.3.1.1 Ecological motivation

The main goal of the ecological motivation is to reduce the overall environmental impact generated by a product. Among other things, emissions, energy consumption and waste production can be reduced by optimising the life cycle of a product unit. The life cycle approach is especially important in this context because the ecological motivation is to lower the overall impact throughout the life cycle of a product. Trade-offs between different phases in the life cycle also have to be taken into account. For example, a product which is very easy to produce may require less energy during its manufacture but consumes a multiple amount of energy during use.

6.3.1.2 Economic motivation

The aim of the economic motivation is to optimise the profits generated by a product during its life cycle. The optimisation of profits for the manufacturer can be achieved in two ways: either directly by lowering costs or creating new revenues for the manufacturer or indirectly through increasing sales by offering products which operate at lower costs. Thus the methods used for assessment of life cycles focus on costs and revenues. As operating costs often exceed acquisition costs, this approach was first considered by users (Mateika 2005). For the manufacturer, the period under consideration for evaluating profitability traditionally ends with the closing of the sale. However, new business models require a more sophisticated approach because the mere sale of a product often fails to generate the required profit (Mateika 2005).

In this case, the manufactured product sometimes becomes solely a vehicle for the sale of services, additional equipment or ancillary materials. One example of this can be found in the printer business in which a large percentage of the manufacturer's turnover is created through the sale of ink cartridges and not through the sale of the printer itself (see Figure 6.5):

Ten ink cartridges were selected on the basis of data provided on the Internet by Hewlett Packard. A black ink cartridge for the HP Deskjet D 2460 contains enough ink to print approximately 150 pages (Hewlett Packard 2007a). Therefore, ten black ink cartridges will thus be needed to print about 1500 pages, an overall output which even an infrequent user will produce during the product's life cycle, given Hewlett Packard's specification of a maximum output per month of 1000 pages (Hewlett Packard 2007b).

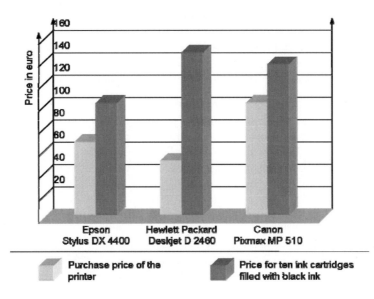

Fig. 6.5: Prices for printers and ink cartridges in Euro. (Internet investigation, 2007)

New legislation such as the take back obligation in the electronic or automobile industry extends the responsibility of the manufacturer right up to the last phase in the product life cycle (EU 2003 WEEE; EU 2000). The manufacturer is forced to take issues such as recycling or remanufacturing into account because economic responsibility for disposal of the product has to be faced. "A high level of waste production and the growing costs of processing and re-integration or burning and disposal of production wastes and obsolete products also generate high company costs" (Züst 1996). The optimisation of life cycles with regard to both economic and ecological goals is a challenge which is discussed in the next paragraph.

6.3.1.3 Synthesis of the economic and ecological motivation

Although the economic and ecological motivations seem to be irreconcilable, it can also be argued that they depend on each other. Without promising economic success or at least neutrality, measures aimed at reducing the environmental impact have little chance of being realised. On the other hand, economically-motivated measures are only sustainable in the long run if they can rely on renewable resources.

A convincing attempt to reconcile these seemingly antagonistic directions has been made by Tomiyama et al. According to Tomiyama et al, the modern mass production paradigm is reaching its limitations. Thus, the "production and consumption volume of artefacts [will have to be reduced] to an adequate and manageable size balanced with natural and social constraints" (Umeda et al. 2000).

This Post Mass Production Paradigm (PMPP) includes two approaches:

- Closing product life cycle loops through maintenance, remanufacturing, component reuse and recycling.
- Dematerialisation: offering services instead of physical products.

Through the "transition from quantitative sufficiency to qualitative satisfaction" (Tomiyama et al. 1996), it is possible to maintain economic growth and high living standards while at the same time reducing the consumption of natural resources. Many ecological objectives such as the economical use of resources or the reuse of components can at the same time be highly profitable. Rising prices for raw materials (see Figure 6.6) diminish the profits of corporations which are greatly dependent on them for the creation of value:

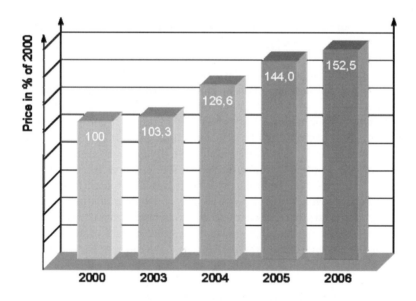

Fig. 6.6: Producer's price of pig iron, crude steel, rolled steel and ferrous alloys in Germany. (Federal Statistical Office Germany 2008a-d)

All major stakeholders are able to profit from measures which reduce the consumption of raw materials: the company through cost reduction, the customer through lower prices and society through the use of fewer resources (Stevels 2000). This two-fold objective to reach economic and ecological goals at the same time is reflected in the requirements for the design of life cycles.

6.3.2 Requirements for the design of life cycles

Due to the fact that companies which exclusively provide services already function on a level of complete dematerialisation, they are not discussed here. The focus is on the producers of manufactured goods or OEMs, regardless of whether the products are sold to businesses or to private customers. An original equipment manufacturer (OEM) buys parts or modules from suppliers to build them into his products (Backhaus 1997). As a result of this position in the supply chain, the OEM is in contact with the customer and is, among other things, responsible for the functional capability and safety of the product. Hence, the OEM is the only party capable of managing the entire life cycle of a product from production till disposal.

Given the objectives described above and the shortcomings of existing models for the design of life cycles (see Chapter 6.2), the requirements for a method to draft and evaluate holistic life cycle concepts are explained in the following.

First of all, the method should be generic. The approach is to design a generic method first which can be modified according to the needs of specific industries or businesses in the future. Due to the strategic and long-term orientation of a new life cycle concept, the current abilities of the organisation and its members are not initially regarded as restrictions. A new life cycle concept will probably require new skills and capabilities anyway. Given its extended horizon, it is possible to hire employees with the required skills, train current employees or outsource certain processes to business partners if necessary. Thus, current abilities are not a criterion during the design of life cycle concepts but do become important in the evaluation of alternative life cycles.

To structure the large number of life cycle measures, they have to be classified. The requirements for this classification result from its use to distinguish different fields of action for the idea creation process. Although these fields of action should closely define the field for which new ideas should be created, they should not be too detailed as this would hinder the idea generation process. Therefore, each field of action represents a basic concept or idea rather than loosing itself in elaborate definitions.

To make life cycle concepts "future-proof", it is essential to include forecasts of the future in the design of life cycles process. However, as it is impossible to predict future developments with certainty, it is highly risky to rely on a single projection. To reduce this risk, multiple scenarios of possible futures are taken into account.

To draft new life cycle concepts which are not restricted by established systems of thought of a certain industry or professional category, is necessary to think "out of the box". This can be achieved through the use of creativity techniques and the participation of outsiders in the process of idea creation. If the concept of a product's life cycle is supposed to be truly holistic, experts from many different professions have to be included right in the early phases of the concept development process. A coherent and successful life cycle concept, which uses the interdependencies between the different phases of the life cycle to the advantage of the overall success of the concept, can only be created if many perspectives are taken into account. As the selection of a new product life cycle concept may have far-reaching consequences for a company, the decision is probably one of corporate-policy and may lead to power struggles within the organisation.By including all departments from the very beginning, not only it is possible to overcome power struggles but also to prevent them from occurring at all.

Decisions about which life cycle concept should be realised need to be based on both an ecological and economical assessment of the concepts. As mentioned previously, both the economic and the ecological motivations are essential for the overall success of a life cycle concept in the long term. A new life cycle concept should therefore be evaluated according to both these criteria. Although a strong customer focus in the product development process is a frequent request, current research shows that the integration of customers in the product development process is not always beneficial (Kohn and Niethammer 2005). Despite this, there are conditions under which the integration of costumers can be crucial to the success of a new product. For these reasons, customers are involved in the assessment of new life cycle concepts. A new life cycle concept can only be successfully implemented with the support of the top management. As it is a decision with far-reaching consequences for the strategic orientation of the organisation, it should be made by the top management anyway. These requirements lead directly to a definition of life cycle concepts.

6.3.3. Definition of life cycle concepts

The aim of the method for the design of life cycle concepts is to describe the life cycle of a physical product. Not only the life cycle of an individual unit, but also the life cycle of a whole product generation is planned. All phases of the life cycle are included, right from the start of the development process up to the disposal of the last remaining product. Flows of material and information are described as being services which are offered together with the physical product. The process of creating the concept has to fulfil the requirements named above. In order to convert the requirements into product concepts, the next chapter starts by classifying life cycle measures.

6.4 Fields of action in life cycle design

The measures to achieve dematerialisation and close the open life cycle loops can be classified into three basic fields of action. Several considerations led to this classification: first of all, there is a large inconsistency in the use of life cycle-related terminology, not only across different industry sectors but also in research.

However, many of the terms have meanings which are largely equivalent to each other (Lindahl et al. 2005). Secondly, the methods and measures for the design and optimisation of life cycles should be easily associated with one of the fields of action. Since the fields of action are also the areas on which the impact of different scenarios is analysed, each field of action should represent one basic concept or idea. This way, they can be used to define the questions used as an input for creativity techniques (see Chapter 6.6.1). The following classification has been developed with these thoughts in mind. All of the life cycle-related terms which are in use can be allocated to one of the three main fields of action:

6.4.1 Material recycling

In everyday life, the term *recycling* is often used as an equivalent of *materials recycling*. In research, the term recycling is sometimes used to address all methods of product recovery (see VDI 2243).

Following Lindahl et al. (Lindahl et al. 2005) and Bullinger et al. (Bullinger et al 2003), material recycling can be defined as the process by which materials otherwise destined for disposal are collected, processed and used for the manufacturing of new products during which the original shape of the material is disintegrated. According to Bullinger et al. (Bullinger et al. 2003), two kinds of material recycling loops can be distinguished:

Waste of production recycling: this waste has the advantage that it is mostly clean and not mixed with other materials. For these reasons it is traditionally widespread.

End-of-life recycling of materials: here there is a problem in that the raw materials are not clean and mixed. However, both forms of material recycling require the disintegration of the original shape of the part or material. Amongst other things, these processes require energy and mostly unmixed materials. The energy and also time and effort involved in complete disassembly can be saved if the shape is maintained, for example, through remanufacturing (Kimura and Suzuki 1996).

6.4.2 Remanufacturing

"Remanufacturing is recycling by manufacturing [...] products from used products [and] is becoming the standard term for the process of restoring used durable products to a "like-new" condition" (Steinhilper 1998). The term *remanufacturing* is used for this field of action for several reasons. It is not only the most widespread term used to address the related issues but also the one which is defined in the same way by many researchers (Lindahl et al. 2005). The "like-new" state which is achieved through remanufacturing may differ. It can be the original condition after manufacture of the machine or a new state as a result of modernising certain components. The latter can be called up-cycling (Bullinger et al. 2003) or upgrading (see Chapter 1.1). Also included in this field of action is reuse and *component cannibalisation*, which can be understood as the use of components or parts which have been extracted from products after at least one life cycle as spare parts or in newly manufactured products (Lindahl et al. 2005). Other terms incorporated into this field of action are reconditioning or *refurbishment*, which can be understood as remanufacturing without any up-cycling (for definitions see Lindahl et al. 2005). Remanufacturing is only profitable and practicable under certain circumstances named by Üffinger (Üffinger 1999):

- High-value product (technically demanding, complex)
- Contains valuable components
- Wear and tear are not extreme
- Contamination is not too severe
- Product does not become technologically obsolete during one life cycle
- No strong dependency on fashion
- Availability at the end of the life cycle
- Modular design preferred (see Chapter 1.1)

These restrictions have to be kept in mind when planning remanufacturing measures. While both material recycling and remanufacturing try to close open loops in the flow of materials, dematerialisation can only be achieved by creating new services.

6.4.3 Services

The creation and extension of services can not only prevent ecological damage through dematerialisation (Umeda et al. 2000) and create additional profits for the manufacturer, but can also support activities from the other two fields of action. By keeping in touch with the user while delivering services, the manufacturer is present as a contact at the time of disposal, which is very important for materials recycling or remanufacturing. If constant maintenance is part of the service, the manufacturer is informed about the whole "life story" of the product. This can be crucial for the success of remanufacturing. Meier (Meier 2003) developed the following model for different levels of services and associated business models as shown in Figure 6.7:

Although this model was originally developed for manufacturing systems and plant engineering, it also offers valuable insights for the producers of other technical goods.

According to Meier (Meier 2003), the first stage includes the obligatory services delivered together with the product such as documentation and warranty. In the second stage, optional services are offered which enable the customer to use the product. The services offered during the third stage secure the operational availability of the product. From the fourth stage onwards, financing is offered and rental and leasing models are available.

Through sale and lease-back, external financial service providers are included. Funding services can also be provided by the manufacturer. One example of this is the automobile industry where large manufacturers even found their own banks, e.g. Mercedes-Benz Bank AG (Mercedes-Benz Bank AG 2008).

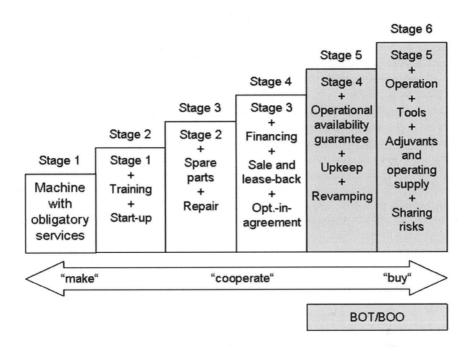

Fig. 6.7: Stage model of services offered along with physical products. (Meier 2003)

An Opt.-in Agreement establishes a direct business connection between the supplier and customer in the case of supplier bankruptcy. By offering maintenance and the optimisation of means of production in the fifth stage, the manufacturer is integrated into the business processes of the customer. In the sixth stage, the manufacturer even provides the staff to operate the machine. If the customer only provides the materials for the production process, the technical equipment supplier turns into a sub-contractor or material supplier. The supplier may even guarantee a certain volume of sales to the customer, thus sharing the risk of a market failure.

The fifth and sixth stages are also referred to as models of BOT or BOO. The BOT (Build-Operate-Transfer) model includes the building of the machine or facility, the operation through the producer and the transfer of ownership of the machine or facility to the customer (Backhaus 1997). In the BOO (Build-Operate-Own) model, the ownership remains with the producer (Backhaus 1997). Spath and Schuster name a third possible configuration which includes leasing: BLOT (Build-Lease-Operate-Transfer) (Spath and Schuster 2004).

As Backhaus states, many - often individual - variations of these models exist (Backhaus 1996). The stages and models described here are therefore to be seen as examples to highlight the multitude of options available for the design of services. Such models offer various advantages not only for the customer but also for the manufacturer. The customer is able to eliminate the financial risks of investment, which is especially important due to today's short product life cycles (Spath and Schuster 2004). At the same time, the customer can save the costs involved in training employees, operation and maintenance (Spath and Schuster 2004). Westkämper claims that "cooperation in partnership of manufacturers and users is an obligatory requisite for maximising the results of any investment" (Westkämper 2006).

For the manufacturer, Spath and Schuster name the following benefits: entering new business segments gives the manufacturer opportunities for growth. It is also possible to establish long-term and intensive customer contact and gain advantages over competitors. As the manufacturer has more experience and competence than the customer, the machines can be operated more efficiently which improves productivity and availability (Spath and Schuster 2004). This knowledge is hard for competitors to copy.

These measures of material recycling, remanufacturing and services are alone merely options for future action. To determine which measures make sense in the context of a specific organisation, it is necessary to take a look at the future. The next chapter is concerned with the question of how such a prognosis can be made and which methods are suitable for the design of life cycle concepts.

6.5 Prognosis

As "today's markets are characterised by a high degree of turbulences" (Niemann 2006), it is necessary to forecast future developments. There are different methods available to create forecasts, ranging from simple linear extrapolations of current trends to more elaborate methods such as the Delphi method or scenario analysis.

Linear extrapolations predict future developments by assuming that the current trend will continue from the past into the future (Specht and Beckmann 1996). Since discontinuities are not taken into account, this method is unfit to predict developments taking place in the more distant future. Specht and Beckmann recommend using them for short- to medium-term prognoses up to five years (Specht and Beckman 1996). For this reason, they can not be used for a method for the design of life cycle concepts which requires long-term prognoses.

Most economic or social developments show non-linear progressions, often shaped like an "S". It has been observed that the path of a technology life cycle often follows such a curve as shown in Figure 6.8:

It was also recognised that new technologies (B) often show poorer performance at the beginning than already-established technologies (A) but soon overtake them. At the same level of performance, the same development effort (Δ E) invested in technology A may create much less improvement in performance (Δ I_A) than in technology B (Δ I_B).

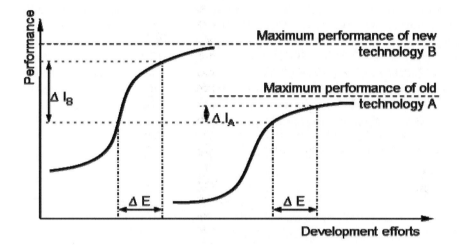

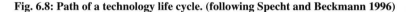

Fig. 6.8: Path of a technology life cycle. (following Specht and Beckmann 1996)

To make predictions based on the assumption that the development follows such a curve is difficult for two reasons. First of all, it is hard to determine where on the S-curve the situation of today is located. Secondly, it is not easy to forecast where the actual border of growth or development is. This makes the S-curve an unreliable tool for making predictions.

A more complex method for forecasting the future is the Delphi method. A panel of experts answers questionnaires in several rounds. After each round, the experts receive a summary of the results plus the anonymised reasoning which led to the judgements (Beck et al. 2000). With this iterative approach, a single consensus is reached. The prognosis is likely to be more accurate if experts from different professions with different experiences and from various institutions are included (Beck et al. 2000). The problem also remains that experts tend to overestimate the importance of their own area of expertise (Gausemeier et al. 1996). Other disadvantages are obvious: such a study is elaborate (Specht and Beckmann 1996) and thus both costly and time-consuming. In addition to that, only one possible development for the future is described. However, to rely on just one single forecast and neglect other possible developments can be dangerous (Gausemeier et al. 1996).

According to Gausemeier et al., forecasting the future faces two big challenges: *multiple futures* and *network thinking* (Gausemeier 1996). With multiple futures, it is possible to take "several future projections into account" (Gausemeier et al. 1996). These multiple futures are used as an input for the idea-creation process (see Chapter 6.6.1) because they are likely to spark more ideas than a single fore-cast. Network thinking is indispensable to make the complexity of modern inter-dependent markets manageable (Gausemeier et al. 1996). A tool to analyse these interconnections and describe several possible futures is scenario analysis. Due to the weaknesses of the other methods and the suitability of scenario analysis for the method for the design of life cycle concepts, it is used here. "A scenario [... is] a description of a future situation, whose occurrence and description of the devel-opment which led from the present to this situation can not be predicted with cer-tainty" (Gausemeier et al. 1996). The use of scenario analysis to describe possible product life cycles of the future results in the formation of life cycle scenarios (Mateika 2005). This includes the prognosis of important influences which may arise during any phase of the life cycle:

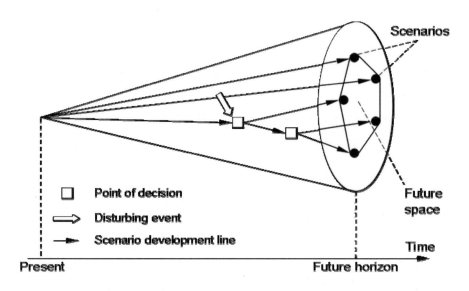

Fig. 6.9: Scenario funnel. (Gausemeier, Fink and Schlacke 1996)

As Figure 6.9 illustrates, scenarios can be seen as being projections from to-day's situation on a future horizon. The development from the present situation to a scenario can be influenced by disturbing events. These may lead to a point where a decision has to be made which influences the occurrence and characteris-tics of the future scenario.

6.5.1 Creation of life cycle scenarios

Following Geschka, Mateika suggests this framework to create life cycle scenarios (Mateika 2005). In a first step, the field under examination is defined as the phases of the life cycle. Information is structured along the phases of the life cycle.

The second step is the identification and structuring of important influencing factors and spheres of influence. The influencing factors describe the state of the spheres of influence, which are substructures with a high significance for the field under examination. Both influencing factors and spheres of influence can be identified using discursive methods (e.g. systematic derivation) or intuitive methods (e.g. brainstorming) (Mateika 2005). All important areas of the environment such as economic, socio-cultural, political, technological and ecological aspects need to be included (Mateika 2005). According to Gausemeier et al., for the most part there are too many influencing factors (Gausemeier 1996). Therefore, *relevant key factors* have to be selected in group discussions which include experts from all phases of the life cycle (Mateika 2005). This process can be supported by comparing the importance of influencing factors in a matrix.

In Table 6.1, a big impact of one influence factor on another is rated with a "3" while little or no impact is indicated by a "0". The impacts are rated by a group of specialists for all phases of the life cycle. (Mateika 2005) The sum of each row is the *active sum* and shows how much a single factor influences the system. The sum of each column is the *passive sum* which indicates how strongly this factor is affected by other factors. Gausemeier et al. differentiate between five types of influencing factors (Gausemeier 1995):

- Impulsive influencing factors (large active, small passive sum): these have a big impact on the system without being influenced by the system itself.
- Dynamic influencing factors (large active and passive sum): these are affected to a large extent by the system and vice versa.
- Reactive influencing factors (small active, large passive sum): these are indicators because the system has a great influence on them while they have no effect on the system.
- Buffer or neutral influencing factors (small or medium active or passive sum): these have little or no impact on the system.

In general, impulsive and dynamic influencing factors are selected as relevant key factors for the life cycle scenario. Along with them, influencing factors should be selected which are highly interdependent with factors from other phases of the life cycle. These factors can facilitate important interdependencies for the success of the product life cycle design (Mateika 2005).

Formulating descriptors and making the projection is the third step in the creation of life cycle scenarios. "A descriptor is a quantitative or qualitative parameter of a key influencing factor." (Mateika 2005) The present state of the descriptors is determined and the state of the future projected. Mateika estimates the probability of occurrence (Mateika 2005) for each state of the future, enabling the likelihood of a scenario to be calculated based on these probabilities. The number of scenarios can be reduced by using their likelihood as a criterion. However, this step of estimating probabilities has been abandoned because such estimates are subjective and not important for the use of the scenarios presented here.

Matrix of Influencing Factors						
on Factor B (column) / Influence of Factor A (row)	Influencing Factor 1	Influencing Factor 2	Influencing Factor 3	⋮	Influencing Factor n	Active Sum
Influencing Factor 1		2	0	…	0	2
Influencing Factor 2	0		3	…	3	6
Influencing Factor 3	1	2		…	2	4
…	…	…	…		…	…
Influencing Factor n	3	0	1	…		4
Passive Sum	4	4	4	…	5	

Table 6.1: Matrix of influencing factors. (Gausemeier et al. 1995)

The projections of the descriptors may contradict each other, making it necessary to assess their consistency in order to exclude non-consistent scenarios. This evaluation can be conducted by filling out a matrix similar to Table 6.2.

By combining the existing variations of factors, all possible combinations of projections are obtained. As the number of combinations rises exponentially with the number of factors, they may become quite numerous. This large number can be reduced by calculating values of consistency and excluding all non-consistent and non-plausible combinations (see Mateika 2005). With the aid of the cluster analysis, the remaining combinations are merged to groups with similar characteristics. This makes it possible to construct consistent and plausible pictures of the future or, in other words, scenarios.

Gausemeier et al. propose the composition of two to four scenarios which should not be either one-sidedly negative or positive (Gausemeier 1995).

Consistency matrix					
with Factor B (column) — Consistency of Factor A (row)	Descriptor of Influencing Factor 1	Descriptor of Influencing Factor 2	Descriptor of Influencing Factor 3	⋮	Descriptor of Influencing Factor n
Descriptor of Influencing Factor 1		1	5	…	3
Descriptor of Influencing Factor 2	1		2	…	0
Descriptor of Influencing Factor 3	5	2		…	1
…	…	…	…		…
Descriptor of Influencing Factor n	3	0	1	…	

5	High Consistency
…	…
1	High Inconsistency

Table 6.2: Consistency matrix. (following Mateika 2005)

These can then be interpreted and conveyed into verbal descriptions. The effects on the environment examined can be determined and concepts for measures and planning drafted (Mateika 2005).

6.5.2 Use of life cycle scenarios

First and foremost, scenarios are used in the method for the design of life cycle concepts to create new ideas. They make it possible to think "out of the box" of every day business. This utilisation is described in detail in the next chapter.

As the ideas created this way form the basis of a new life cycle concept, it is essential to keep track of the development in reality. If none of the underlying scenarios of a life cycle concept actually become reality, the life cycle concept runs the risk of failing.

Once a life cycle concept has been selected, the ideas characterising the concept are traced back to the features of the scenarios which the idea aims at. It is not the occurrence of whole scenarios but rather of developments which is crucial for the success of the ideas used. This reduces the effort required for monitoring. Furthermore, it is desirable to detect such developments as early as possible. This makes it easier and cheaper to readjust the life cycle concept.

Ansoff proposes the concept of weak signals to enable such an early detection of discontinuities which are "significant deviations of the actual development from the predicted development" (Zeller 2003). Zeller calls attention to the inconsistent and vague definitions of weak signals (Zeller 2003) and names the following characteristics of weak signals:

They anticipate future developments and are
- qualitative,
- ambiguous,
- fragmentary,
- have no wide diffusion nor
- clear cause-effect relationships.

For these reasons, weak signals can be interpreted in various ways and have to be pieced together from different information fragments (Zeller 2003). These features of weak signals are the cause of the difficulties of putting the concept into operation. A promising attempt to realise this operationalisation through data mining has been made by Zeller (Zeller 2003). It seems inadvisable to rely on weak signals because of these difficulties. A well-established method of marketing research in order to observe current developments in markets is the use of panels. More details are given in Chapter 6.7.3.

Once deviations have been detected by weak signals or actual developments, a quick reaction is necessary. Gausemeier et al. suggest a process for the development of "future-robust" strategies to enable such quick reactions to detected deviations (Gausemeier et al. 1995).

In a first step, the implications of the occurrence of a certain scenario are assessed. This may also include deviations (Gausemeier et al. 1995). In a second step, contingency plans are developed on the basis of the chances and risks of the scenarios. In the final step of "robust" planning, several contingency plans are combined to form a strategy offering an appropriate reaction to the occurrence of several scenarios (Gausemeier et al. 1995).

Independent from these additional uses of scenarios is the main utilisation as an input for the development of life cycle concepts.

6.6 Development of life cycle concepts

The aim of this section is to design a new life cycle concept. Creativity is indispensable for developing innovative ways to close life cycle loops and plan new services. However, process of idea generation needs to be shortened and directed towards feasible solutions. This can be achieved by using the scenarios described above as a framework.

6.6.1 Idea generation

To invent new and even radical solutions either to close open product life cycle loops or create new services, it is necessary to look at the structures and their associated industrial problems from a different point of view. As specialists possess considerable knowledge in their field of expertise, they are used thinking in a certain way along the lines of the status quo. It is hard for them to question basic "rules" of their subject, even if those are simply a result of historic developments. A homogeneous group consisting of experts from only one profession will thus most likely create ideas which are only references to existing ideas (Schnetzler 2004). Outsiders have the huge advantage that they do not even know these rules and are thus free of restricting themselves to well-tried solutions. A mix of experts from inside the organisation and outsiders is the ideal for creating ideas (Schnetzler 2004). These outsiders can be experts from completely different areas, students, customers, people with various cultural backgrounds and even teenagers. To include teenagers in the search for new ideas is practised systematically by the Brain-Store AG, a successful service provider for the creation of ideas (Schnetzler 2004).

Another approach stresses the *destructive* element of all creativity techniques which is supposed to destroy conventional patterns of thought to enable the *constructive* part of finding new combinations to emerge (Eversheim and Schuh 1999).

To create ideas efficiently, there has to be a clear goal and the problem must well-outlined (Ophey 2005). To achieve this, the fields of action and the scenarios developed in the preceding paragraph are used to find ways to close product life cycle loops or invent new services (Table 6.3). In addition to the three fields of action from the life cycle approach, customers are added as a fourth field of action to add another dimension to the idea creation process which includes their points of view. The scenarios and the fields of action not only provide a frame for the search but also offer a much more important aspect, i.e. a stimulus for the creation of ideas. For the fields of action, the basic thought behind the concepts of *material recycling*, *remanufacturing* and *services* inspire ideas. The scenarios are provided in the form of verbal descriptions of possible futures and allow the participants to let their thoughts wander freely, unconstrained by today's restrictions. This also enables problems which were previously unnoticed to be detected.

Matrix for Idea Generation				
Scenarios Fields of Action	Scenario 1	Scenario 2	...	Scenario n
Material recycling	How could material re-cycling work if Scenario 1 occurs?	How could material re-cycling work if Scenario 2 occurs?	...	
Remanufacturing	...			
Services				
Customers	What kind of product would customers need if Scenario 1 occurs?			

Table 6.3: Matrix for idea generation.

For each field of the matrix, new ideas are created using such techniques as brainstorming, method 635, synectics or bionics. A brief outline of the methods named in Table 6-4 and Table 6.5 is given below.

Brainstorming is one of the most widespread creative techniques used (Schnetzler 2004). A question is asked by the moderator and the participants articulate their ideas which are recorded. It is important to note that criticism is not allowed during this process and all ideas, however irrelevant they may seem at the time, are written down (Schnetzler 2004). A great advantage of brainstorming is that participants can inspire each other. The quality of the brainstorming depends to a great extent upon the moderator (Schnetzler 2004).

Method 635 belongs to the brainwriting techniques. The participants write down their ideas and hand them over to the next person in the group, enabling them to inspire each other. By leaving little time for this process, it can be avoided that the participants think too much about their ideas (Schnetzler 2004).

Creativity Technique	Principle	Team size	Steps	Function of moderator	Documentation	Special Restrictions	Goal
Brain-storming	Asso-ciativity	5-12	1. Problem definition outside the group 2. Information of team members a few day prior to the session if necessary 3. Sessions with unconstrained course 4. Evaluation through experts, first selection through participants if necessary	- Prevent critique - Stimulate discussion - Discontinue discussion if no ideas are spoken out	Pin board, audio tape, video, flip-charts	- No critique allowed - Enhance expressed ideas further	- Maximise number of ideas - Make ideas as extravagant as possible
Method 635	Asso-ciativity	6	1. Problem definition outside the group 2. Internal agreement of the group concerning the contour of the problem 3. Every participant writes down three ideas 4. Sheets are passed: every participant takes a stand on her/his neighbours ideas and develops new ideas 5. Sheets are passed another four times 6. Systematic evaluation	- Foster an agreement on problem definition - Time- and documentation management	Form	- Participants do not need to know each other	- Produce precisely articulated ideas - Restrict discussions Ensure participation of all members

Table 6.4: Creativity techniques. (Eversheim and Schuh 1999)

Creativity Technique	Principle	Team size	Steps	Function of moderator	Documentation	Special Restrictions	Goal
Synectics	Associativity	5-7	1. Train synectics 2. Problem definition outside the group 3. Reaching a consensus on the understanding of the problem 4. Selection of spontaneous solutions 5. Gaining "distance" to the problem, "alienation" of the problem, personal analogy 6. Establishing new associativity, analogies 7. "force-fit": connecting associativity and analogies with the problem 8. Development of concrete solutions	- Training - Foster an agreement on problem definition - Selection of associativities and analogies for "alienation" - Overcome inhibitions of "alienation" - Determine when to switch from "alienation" to "force-fit" Describe solutions	Pin board, audio tape, video, flip-charts	- No un-selfconscious use without training - Deliberate stimulation of emotions - Avoid critique and internal alienation	- Make ideas as unconventional as possible - Produce rationally uncontrolled ideas - Give up attitudes, dissolve existing patterns of thought
Bionics	Analogy	-	1. Problem definition 2. Systematic search for analogous solutions in nature 3. Theoretical foundation of the evolutionary process in nature 4. Examination of transferability 5. Systematic reconstruction of evolutionary process if necessary	No moderator	Conventional forms of documentation	None	Reproduce natural process of evolution and use the power of selection through nature

Table 6.5: Creativity techniques. (Eversheim and Schuh 1999)

"Synectics defines the creative process as the mental activity in problem-stating, problem-solving situations where artistic or technical inventions are the result" (Gordon 1961). Gordon stresses the importance of distinguishing between the two steps of *problem-stating* and *problem-solving*. By stating the problem, an understanding of the actual problem is reached: the strange is made familiar (Gordon 1961). In the next step of problem-solving, the familiar is made strange (alienation): the "conscious attempt to achieve a new look at the same old world" (Gordon 1961) is made. Analogies are used to achieve this alienation. In a final step these analogies are force-fitted with the problem to create novel ideas and solutions.

Bionics "systematically studies the technical implementation and use of constructions, processes and principles of evolution of biological systems" (Neumann, cited in Hill 2005). Solutions created by nature are not only environmentally sound, but also reach a maximum of functionality and reliability with a minimum of material and energy (Hill 2005). For these reasons, similar systems in nature are analysed and their structure is abstracted in order to detect the underlying principle. These principles can be varied or combined to fulfil the requirements for technical solutions (Hill 2005). For more creativity techniques, see Schnetzler (Schnetzler 2004) or Bullinger et al. (Bullinger et al. 2003).

Matrix for the Collection of Ideas				
Scenarios Fields Of Action	Scenario 1	Scenario 2	...	Scenario n
Material Recycling	- Idea A - Idea B	- Idea C - Idea D	...	- Idea B
Remanufacturing	...			
Services				
Customers				

Table 6.6: Matrix for the collection of ideas.

The ideas created using these methods can be plotted in Table 6.6. Collecting ideas in this way has the advantage that it is easier to determine similar ideas which were developed in the same field of action but for different scenarios. It is not advisable to generate life cycle concepts directly out of this structured but large number of ideas. The number of possible combinations of ideas is just too extensive to form a life cycle concept and likely to get out of hand. The ideas need to be refined and selected before taking this step.

6.6.2 Idea assessment

The multitude of ideas created in the previous step has to be reduced to a manageable amount. Two criteria are taken into consideration here: firstly the frequency of appearance and secondly the value or benefit of an idea.

6.6.2.1 Frequency of appearance

A group of experts from each field of action select the most promising ideas and work out the basic principles behind each idea. During this process the ideas are automatically refined: intuitive and general ideas become more closely defined and are concretised. This way it is possible to find out if certain ideas appear several times in different scenarios. The recurring ideas are then ranked according to the frequency of appearance (Figure 6.10).

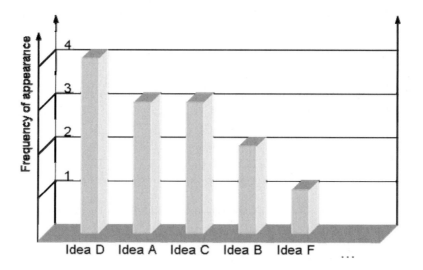

Fig. 6.10: Ranking of ideas according to frequency of appearance.

An idea which recurs several times fits multiple future developments and is thus more *future-proof* than an idea which is only likely to be successful if one specific scenario occurs. This method is equivalent to Gausemeier's approach to form strategies which are *future-robust* out of several plans made for single scenarios (Gausemeier 1996). To rely on several scenarios also reflects another life cycle-related thought: products should be long-living, multifunctional and flexible in terms of alterations (see Tomiyama 1996). Naturally this approach favours platform concepts and modular features (see Westkämper 2003).

By using the frequency of appearance as a criterion, niche markets which are too specific do not become the goal of a life cycle concept. This reduces commitments to such - mostly small - markets which could easily vanish, although it is still possible to enter these markets on the basis of a standard platform combined with modular features.

However, it is impossible to draw conclusions regarding the value of an idea just from the number of appearances. Ideas which offer only little improvement may still appear several times. Thus it is essential to assess the anticipated benefit of an idea.

6.6.2.2 Value benefit analysis

To assess the true value of an idea, it is necessary to take both qualitative and quantitative measures into account. Quantitative measures are usually preferred because of their easy comparability. However, many qualitative measures are hard to transform into quantitative measures. One method of converting qualitative into quantitative measures is the value benefit analysis which is described in the following.

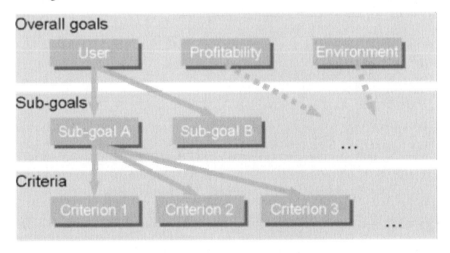

Fig. 6.11: Hierarchy of goals. (Hoffmeister 2000),(Pahl et al. 2007)

The first step of an evaluation is to define goals. According to Pahl et al. (Pahl et al. 2007), these goals have to meet the following requirements:

They should cover the requirements and constraints as completely as possible. At the same time, the individual goals should be as free of interdependencies as possible, a demand also made by Hoffmeister (Hoffmeister 2000). It also has to be possible to determine the actual characteristics of the item to be valued with an appropriate effort. Both Pahl et al. (Pahl et al. 2007) and Hoffmeister (Hoffmeister 2000) build up a similar hierarchy of goals. Pahl et al. separate the goals horizontally into different target fields. In accordance with Chapter 6.7, three target fields are proposed: the user or customer, profitability and the environment. Vertically, these overall goals can be divided into sub-goals and even further into single criteria (Figure 6.11). The question of how detailed the hierarchy should be depends on the amount of time available and the number of ideas to be assessed. If a large quantity of ideas needs to be considered, it may prove more practicable to evaluate the ideas first only according to the three overall goals in order to make a preliminary selection. In a second step, the ideas can then be assessed in more detail with regard to sub-goals or criteria.

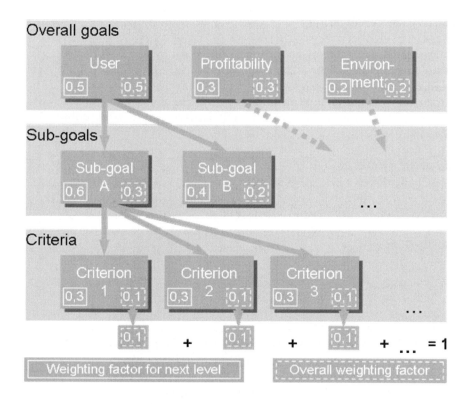

Fig. 6.12: Weighting of goals. (Pahl et al 2007), (Hoffmeister 2000)

However, not only is it crucial for an idea to achieve a particular objective to a certain degree, the relevance of that objective compared to others also needs to be assessed. The importance of the three overall goals has to be weighted against each other in the same way as the sub-goals and criteria need to be weighted according to their importance for the next higher level (Pahl et al. 2007). By multiplying the weighting factor of a level with the factor of the next lower level, the weighting factor for the overall goal (overall weighting factor) of the lower level can be calculated. The sum of all weighting factors on one level must add up to 1.0 (see Figure 6.12). Only then is it possible to evaluate the ideas themselves.

On a scale ranging from 0 to 10, the evaluator assesses the degree of performance of each idea with regard to a criterion (Pahl et al. 2007). The same process can be used on the two other levels, for example for a pre-selection. The fact that these assessments are estimates and that they are subjective may not be neglected. If the evaluation is carried out on the highly abstract level of overall goals, these estimates are more likely to be highly subjective than on the lowest level where the definitions of the criteria are more concrete.

The total value benefit of an idea is calculated by multiplying the score of the criteria and sub-goals or overall goals with the weighting factor of the criteria, sub-goals or overall goal and then calculating the sum of all items on the respective level (Hoffmeister 2000). A selection on the basis of these values can be made in two ways: either a specific value is set as a minimum or a certain percentage or number of the ideas with the highest score is chosen. The selected ideas form the basis for the next step, the creation of a life cycle concept.

6.3.3 Creation of life cycle concepts

Due to the amount of possible configurations, it is a difficult task to design and assess possible life cycle concepts. To support this process, the interdependency matrix is proposed as a tool. The ideas for the three fields of action have various connections and trade-offs amongst each other. It is important to take these potential conflicts and positive trade-offs into account when creating life cycle concepts. "The planning of the product and the corresponding life cycle of the product cannot be separated from each other" (Kölscheid 1999). The interdependencies between each hypothetical combination of ideas can be recorded in the following Table 6.7.

The ideal life cycle concept is characterised by many positive interdependencies through which the intended effects of every realised idea are strengthened. The use of the interdependency matrix prevents conflicts in the implementation of the life cycle concept and generates more harmonic life cycle concepts.

The interdependency matrix can be used in two ways: if there are only few ideas, it is possible to compute the sum of every theoretically possible configuration of ideas. In this case the configurations with the highest sums are selected. If there are too many ideas to compute the outcome of every configuration, the interdependency matrix is used to assess the coherence of life cycle concepts developed by a group of experts. The result is then used to select the most harmonic life cycle concepts. Below is an example of a hypothetical life cycle concept including the ideas A, D and F (see marked fields):

$$\sum (A, D, F) = -2 + 2 - 1 = -1 \qquad [6\text{-}1]$$

The final outcome of this stage should be the creation of a maximum of three life cycle concepts. Two considerations lead to this restriction. The customer plays an important role in the subsequent stage of assessment and selection. As Schnetzler states, no more than three options should be presented to the customer (Schnetzler 2004). While the other two tools of assessment - the analysis of costs and revenues and the life cycle analysis - have no such immanent restrictions, they are still costly and time-consuming. Therefore, their use is limited to the development of promising life cycle concepts.

Another reason is that a life cycle concept consists of more than just a selection of ideas. These ideas merely form the basis of the life cycle concept by providing a framework for a more detailed characterisation. This characterisation is not only needed for the assessment but also for the implementation of the life cycle concept. For these reasons, a life cycle concept is a description of all phases of the product life cycle (see Chapter 2.2), including all flows of material (Figure 6.13):

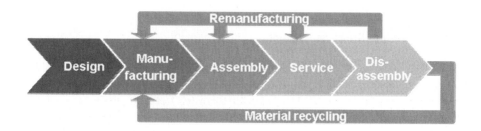
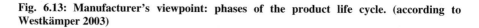

Fig. 6.13: Manufacturer's viewpoint: phases of the product life cycle. (according to Westkämper 2003)

The efforts required to design and engineer new or changed products have to be estimated. The extent to which already existing modules can be used or whether new ones need to be developed gives an initial idea. The new processes can be described based on the current manufacturing and assembly process. Among others, these include the material, the processing of this material and the necessary staff. This highlights the investments required to implement the life cycle concept. The services making up part of the life cycle concept are specified to make it possible to appraise the cost of building up new services or changing service processes already in existence (see Chapter 1.1 and 5.3). The required personnel, equipment and related costs are also determined. The last phase of disassembly and disposal includes material recycling, remanufacturing and the disposal of waste.

Interdependency Matrix												
Fields of Action		Material Recycling			Remanufacturing			Services			Customers	
	Idea	A	B	C	D	E	F	...				
Material Recycling	A	▨	+ +	-	- -		+ +					
	B	+ +	▨	0								
	C	-	0	▨								
Remanufacturing	D	- -			▨		-					
	E					▨						
	F	+ +			-		▨					
Services	...	...						▨				
									▨			
										▨		
Customers											▨	
												▨
													▨

+ + +	Large positive interdependency = +3
+ +	Medium positive interdependency = +2
+	Small positive interdependency = +1
0	No interdependency = 0
-	Small negative interdependency = -1
- -	Medium negative interdependency = -2
- - -	Large negative interdependency = -3

Table 6.7: Interdependency matrix.

All modules or parts are listed together with a specification of their treatment at the end of the product's life cycle. For all these phases, the logistics and flows of material as well as the interfaces to the other phases are described. This comprehensive description of the product and its life cycle can also be considered for the purposes of life cycle modelling (see Chapter 2).

Obviously, these specifications show the life cycle concept from the manufacturer's point of view. However, the customer's viewpoint is also included in the evaluation. A description for users focuses on the manufacturer's offers, with the basic part of this description containing the product specification supplemented by the services offered and concepts for disposal. In this context, it may be necessary to heighten the customer's awareness of actual costs created by the use and disposal of the product. Once these descriptions of the life cycle concepts have been compiled, the life cycle concepts can be evaluated.

6.7 Life cycle concept assessment and selection

There are numerous criteria and methods for evaluating life cycle concepts. Three methods are presented in the following chapter. The two motivations for designing life cycles are reflected in the evaluation criteria. From the ecological point of view, the environmental impact of a life cycle concept is the decisive factor when selecting a life cycle design. To estimate environmental impact, the method of life cycle assessment (LCA) is used.

On the other hand, the economical motivation stresses the importance of considering costs and revenues when making the selection. Another requirement for the economic success of a life cycle concept is the usefulness of the concept to the customer. To develop a truly holistic life cycle concept, all these criteria have to be taken into account.

The question remains as to whether it is necessary to analyse each life cycle concept regarding all three aspects. The manufacturer decides on the method of choice in order to make the decision. By using a single method for the selection of a concept, the manufacturer is able to reduce the time, effort and costs involved in making the assessment. In a second step, the two other methods are implemented to evaluate the selected concept regarding the other aspects. This way, only one method has to be applied three times and the other two only once as a "safety check" or to further refine the concept. The effort can be reduced even further by relying on one method only and neglecting the other methods. The risk of selecting a life cycle concept this way varies depending on the industry and the importance of the product to the manufacturer. The most important basis for decisions in the majority of organisations are the costs and revenues of the selectable alternatives.

6.7.1 Costs and revenues throughout the life cycle

A number of tools and methods are available to analyse costs for the customer. These life cycle cost concepts (often called Life Cycle Costing or LCC), which were developed from the 30's onwards of the last century, focus on the running expenses for long-term investments made by the state and later also by the private sector (see Perlewitz 1998).

The perspective of the manufacturer or service provider has been less investigated. There are many challenges, especially when costs and revenues are supposed to be analysed in the early phases of the product planning process. Large uncertainties are connected to the technical feasibility and economic success of innovations and it is difficult to predict sales volume or prices in the future. Due to long planning intervals, these projections become even less reliable. (Hahner 2000)

It is important that these restrictions underlining the importance of ongoing control of the progress of the planning process and observation of changes in the environment are borne in mind. Given that there is an existing product comparable to the one to be developed, the existing product costs and revenues can be analysed. If there is no such product, predictions of future costs and revenues become more unreliable. However, they remain important for the control of the actual costs and revenues. Elaborate discussions of the analysis of costs throughout the life cycle are available (Schmidt 2000) see also: (Senti 1994), (Osten-Sacken 1999), (Zehbold 1996). But as Kobayashi stated in 2003, "some cost estimation methods for the early phases of design have been reported; however, they are inadequate" (Kobayashi 2003). Given the early stage of product development, the uncertainties described beforehand and the fact that existing models give no systematic derivation and justification of revenues for the manufacturer (Mateika 2005; see also Chapter 1.1), the model developed by Mateika in 2005 is used here as a framework.

Fig. 6.14: Phases of the product life cycle. (following Mateika 2005)

There are two basic cost concepts used to examine costs throughout the life cycle. The first one is based on value and describes the consumption of value in the production of goods and services. The second one is cash-based and views costs as payouts. Since life cycle-costing is long-term orientated and has the character of capital budgeting, this view is preferred by most scholars (Schmidt 2000), (Mateika 2005), (Hahner 2000). Mateika allocates all costs and revenues throughout the life cycle to three basic phases of the product life cycle: development and manufacturing phase, market and service phase and disposal phase, as shown in Figure 6.14. Following Mateika all costs for the manufacturer in the whole life cycle can be allocated to one of these phases in the following form (Mateika 2005):

$$C_{tm} = C_{dp} + C_{mp} + C_{dp} \qquad [5\text{-}1]$$

C_{tm} = Total life cycle costs for the manufacturer [€]
C_{dp} = Costs of the development and manufacturing phase [€]
C_{mp} = Costs of the market and service phase [€]
C_{dp} = Costs of the disposal phase [€]

All these cost pools are described in the following. (Figure 6.15)

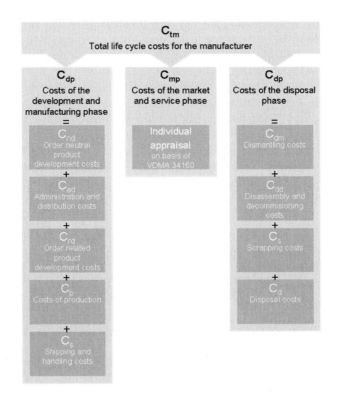

Fig. 6.15: Cost structure.

6.7.1.1 Development and manufacturing phase

In this phase, the processing of an order requires the activities of sales and marketing, design, job preparation, manufacturing and assembly and, preceding that, order-neutral product development (Mateika 2005). Order-neutral product development costs are created independently from any orders before a product can be offered to customers. Administration costs accrue while a product is being developed. Once a development has been completed, distribution costs are created before the product is actually produced, which generates production costs. After production, the product has to be handled and shipped to the customer, sometimes at the expense of the manufacturer. Following Mateika, the costs during this phase consist of (Mateika 2005):

$$C_{dp} = C_{nd} + C_{ad} + C_{dr} + C_{p} + C_{s} \qquad [5\text{-}2]$$

C_{dp}	=	Costs of the development and manufacturing phase [€]
C_{nd}	=	Order-neutral product development costs [€]
C_{ad}	=	Administration and distribution costs [€]
C_{rd}	=	Order-related product development costs [€]
C_{p}	=	Costs of production [€]
C_{s}	=	Shipping and handling costs [€]

6.7.1.2 Order-neutral product development

This phase includes the clarification of requirements, the determination of functions, finding of solutions, structuring of modules, design of modules and products and the design of instructions for use (VDI 2221 1993). Prototypes may be built during the product design phase, creating costs for material and machine utilisation. At this point, Mateika faces the problem that calculatory costs are included in the costs for machine utilisation. This contradicts the presumption that costs and revenues are created at the same time as cash in- and outflows (Mateika 2005). To solve this problem, Mateika proposes the following assumption: if the machines are utilised by an external supplier to produce the parts, this supplier would include the machine overhead rate in his calculation and bill the user of this service accordingly.

The sum of the costs is then divided by the number of benefiting products to allocate these costs directly (Mateika 2005). This view can be problematic as Schmidt points out: the knowledge gained can not only be used for current, but also for future projects, thus making the attribution to a certain number of products difficult (Schmidt 2000). For this reason, the costs for basic research are sometimes viewed as an investment (Schmidt 2000). However, as there is no practicable alternative, Schmidt advocates the method used here to spread the costs over all benefiting products in the form of a "royalty" (Schmidt 2000).

In addition to Mateika, Schmidt drafts an elaborate model which takes the chronological sequence of manufactured products into account that will not be discussed here (Schmidt 2000). The costs for the design of services are added here to Mateika's formula (Mateika 2005). Due to the immaterial character of services, these costs consist mainly of personnel costs.

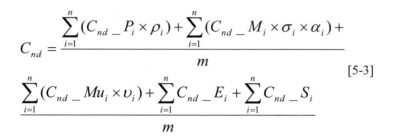

$$[5\text{-}3]$$

C_{nd}	=	Order-neutral product development costs [€]
i	=	Stage of development
n	=	Total number of stages of development
P	=	Personnel [h]
ρ	=	Personnel cost rate [€/h]
M	=	Amount of material [kg]
σ	=	Price of material [€/kg]
α	=	Material overhead rate
Mu	=	Machine utilisation [h]
υ	=	Machine utilisation overhead rate [€/h]
E	=	Costs for external services [€]
S	=	Personnel costs for service development [€]
m	=	Number of benefiting products

According to VDMA, the machine utilisation overhead rate includes costs for cost-accounting depreciation, maintenance, occupancy, energy and tools together with miscellaneous costs such as costs for operation supply items or programming (VDMA 2002):

$$\upsilon = \frac{C_{ad} + C_{ii} + C_{ma} + C_o + C_e + C_t + C_{mi}}{t_{as}} \qquad [5\text{-}4]$$

υ	=	Machine utilisation overhead rate [€/h]
C_{ad}	=	Cost-accounting depreciation [€]
C_{ii}	=	Imputed interest [€]
C_{ma}	=	Costs of maintenance [€]
C_o	=	Occupancy costs [€]
C_e	=	Energy costs [€]
C_t	=	Tool costs [€]
C_{mi}	=	Miscellaneous costs [€]
t_{as}	=	Annual serviceable time [h]

6.7.1.3 Costs of administration and distribution

Costs of overall administration and distribution are often jointly assessed. Thus, Mateika integrates them here (Mateika 2005):

$$C_{ad} = C_{aor} + C_d \qquad [5\text{-}5]$$

C_{ad}	=	Administration and distribution costs [€]
C_{aor}	=	Administration overhead costs [€]
C_d	=	Distribution costs [€]

The process of distribution can be subdivided into initiation, tender preparation, negotiation and settlement of the project (Backhaus 1997). Distribution costs can thus be expressed as (Mateika 2005):

$$C_d = C_{di} + C_{dtp} + C_{dn} + C_{dsp} + C_{dor} + C_{dsd} \qquad [5\text{-}6]$$

C_d	=	Distribution costs [€]
C_{di}	=	Initiation costs [€]

$$C_{dtp} \quad = \quad \text{Tender preparation costs [€]}$$
$$C_{dn} \quad = \quad \text{Negotiation costs [€]}$$
$$C_{dsp} \quad = \quad \text{Settlement of the project costs [€]}$$
$$C_{dor} \quad = \quad \text{Distribution overhead costs [€]}$$
$$C_{dsd} \quad = \quad \text{Special direct costs of distribution [€]}$$

As distribution costs are mainly personnel costs, the equation [5] above is transformed by Mateika into (Mateika 2005):

$$C_d = \sum_{i=1}^{n} C_{di}\_P_i \times \rho_i + \sum_{i=1}^{n} C_{dtp}\_P_i \times \rho_i + \sum_{i=1}^{n} C_{dn}\_P_i \times \rho_i +$$

$$\sum_{i=1}^{n} C_{dsp}\_P_i \times \rho_i + C_{dor} + C_{dsd}$$

[5-7]

i	=	Distribution step
n	=	Total number of distribution steps
P	=	Personnel [h]
ρ	=	Personnel cost rate [€/h]

6.7.1.4 Order-related product development costs

Durable means of production and other capital-intensive goods are often subject to life cycle considerations. These products are often customised according to the special needs of the customer, thus creating order-related product development costs. These costs can be calculated along the lines of order-neutral development costs without dividing them by the number of benefiting products (Mateika 2005) as they are already clearly assignable to a specific product:

$$C_{rd} = \sum_{i=1}^{n} (C_{rd}\_P_i \times \rho_i) + \sum_{i=1}^{n} (C_{rd}\_M_i \times \sigma_i \times \alpha_i) +$$

$$\sum_{i=1}^{n} (C_{rd}\_Mu_i \times \upsilon_i) + \sum_{i=1}^{n} C_{rd}\_E_i$$

[5-8]

C_{rd}	=	Order-related product development costs [€]
i	=	Stage of development
n	=	Total number of stages of development

P	=	Personnel [h]
ρ	=	Personnel cost rate [€/h]
M	=	Amount of material [kg]
σ	=	Price of material [€/kg]
α	=	Material overhead rate
Mu	=	Machine utilisation [h]
υ	=	Machine utilisation overhead rate [€/h]
E	=	Costs for external services [€]

6.7.1.5 Costs of production

The costs of production include the costs for job preparation, manufacture, assembly and quality assurance and are expressed as (Mateika 2005):

$$C_p = C_{jp} + C_m + C_a + C_{qa} \qquad [5\text{-}9]$$

C_p	=	Costs of production [€]
C_{jp}	=	Costs of job preparation [€]
C_m	=	Cost of manufacture [€]
C_a	=	Assembly costs [€]
C_{qa}	=	Costs of quality assurance [€]

All tasks of planning, government and supervision of manufacturing and assembly belong to the field of responsibility of job preparation. Working steps are the planning of equipment, of the manufacturing programme and operating sequence. Given the large proportion of human labour in job preparation, the costs can be calculated using (Mateika 2005):

$$C_{jp} = \sum_{i=1}^{n}(C_{jp} \_ P_i \times \rho_i) \qquad [5\text{-}10]$$

C_{jp}	=	Costs of job preparation [€]
i	=	Stage of development
n	=	Total number of stages of development
P	=	Personnel [h]
ρ	=	Personnel cost rate [€/h]

Manufacturing is the creation of components with specified material properties and dimensions as well as the joining of such components to create products (Czichos et al. 2004). The costs of manufacture include the material, personnel, machine, waste, special direct costs and costs for external services. The theory behind material, personnel and machine costs is explained above. The waste costs are calculated from the quantity of a certain type of waste W_j and the specific disposal cost rate γ_j (Mateika 2005):

$$C_m = \sum_{i=1}^{n}(C_m\_P_i \times \rho_i) + \sum_{i=1}^{n}(C_m\_M_i \times \sigma_i \times \alpha_i) +$$

$$\sum_{i=1}^{n}(C_m\_Mu_i \times \upsilon_i) + \sum_{j=1}^{n}(C_m\_A_j \times \gamma_j) + E_{ms} + E_{mes}$$

[5-11]

C_m	=	Cost of manufacture [€]
i	=	Manufacturing step
n	=	Total number of manufacturing steps
P	=	Personnel [h]
ρ	=	Personnel cost rate [€/h]
M	=	Amount of material [kg]
σ	=	Price of material [€/kg]
α	=	Material overhead rate
Mu	=	Machine utilisation [h]
υ	=	Machine utilisation overhead rate [€/h]
W_j	=	Quantity of waste type j [kg]
γ_j	=	Disposal cost rate type j [€/kg]
E_{ms}	=	Special direct costs of manufacturing [€]
E_{mes}	=	Costs for external services of manufacturing [€]

"Assembly is the mounting of parts and/or assemblies to create a manufactured item or assembly on a higher level of manufacturing" (VDI 1990). Assembly costs are calculated in the same way as for the costs of manufacture except for the fact that there are no significant amounts of waste involved. The costs for the disposal of waste can thus be disregarded (Mateika 2005). While Mateika does not take machine costs into account due to his area of application which is the machine and plant engineering industry, they are included here to keep the model universal:

$$C_a = \sum_{i=1}^{n}(C_a\_P_i \times \rho_i) + \sum_{i=1}^{n}(C_a\_M_i \times \sigma_i \times \alpha_i) +$$

$$\sum_{i=1}^{n}(C_a\_Mu_i \times \upsilon_i) + E_{as} + E_{aes}$$

[5-12]

Ca	=	Assembly costs [€]
i	=	Assembly step
n	=	Total number of assembly steps
P	=	Personnel [h]
ρ	=	Personnel cost rate [€/h]
M	=	Amount of material [kg]
σ	=	Price of material [€/kg]
α	=	Material overhead rate
Mu	=	Machine utilisation [h]
υ	=	Machine utilisation overhead rate [€/h]
E$_{as}$	=	Special direct costs of assembly [€]
E$_{aes}$	=	Costs for external services of assembly [€]

Quality assurance is defined as "all planned and systematic actions realised within the quality management system, which are presented in a way necessary to create sufficient assurance that an entity fulfils the requirements for quality" (DIN EN ISO 8402 1995). The quality assurance costs are included in the quality costs and consist of costs for the management of quality assurance as well as failure prevention costs. Quality costs also include the testing costs, which are mainly personnel costs, costs for measurement instruments and failure costs. Failure costs include the costs of culls and rectification (Mateika 2005):

$$C_q = C_{qa} + C_{qt} + C_{qf}$$

[5-13]

C$_q$	=	Quality costs [€]
C$_{qa}$	=	Quality assurance costs [€]
C$_{qt}$	=	Testing costs [€]
C$_{qf}$	=	Failure costs [€]

Quality costs can thus be calculated by:

$$C_q = \sum_{i=1}^{n}(C_{qa} \_ P_i \times \rho_i) + \sum_{i=1}^{n}(C_{qa} \_ P_i \times \rho_i) + \sum_{i=1}^{n}(C_{qt} \_ Mu_i \times \upsilon_i)$$

$$+ \sum_{j=1}^{n}(N_{cj} \times V_j) + C_{qr}$$

[5-14]

C_q	=	Quality costs [€]
C_{qa}	=	Quality assurance costs [€]
i	=	Step of quality assurance
n	=	Total number of steps of quality assurance
P	=	Personnel [h]
ρ	=	Personnel cost rate [€/h]
C_{qt}	=	Testing costs [€]
Mu	=	Machine utilisation [h]
υ	=	Machine utilisation overhead rate [€/h]
N_{cj}	=	Number of culls
V_j	=	Value of a cull [€]
C_{qr}	=	Costs for rectification of rejects [€]

6.7.1.6 Shipping and handling costs

The last steps of the development and manufacturing phase are the shipping and putting into operation. Although shipping and handling costs are not always paid by the manufacturer, they are included here because such services can be an additional source of revenue. The costs for shipping and handling include packing costs which are made up of packaging material and personnel costs. Carrying costs have to be considered together with installation costs (Mateika 2005):

$$C_s = C_{sp} + C_{sc} + C_{si} \qquad [5-15]$$

C_s	Shipping and handling costs [€]
C_{sp}	Packing costs [€]
C_{sc}	Carrying costs [€]
C_{si}	Installation costs [€]

6.7.1.7 Revenues from the development and manufacturing phase

In general, revenues are only created through the sale of products to the customer on completion of the development and manufacturing phase. The sale itself marks the beginning of the market and service phase. Nevertheless, subsidies, money from external research funds and royalties for licenses can be regarded as revenues created during that phase (Mateika 2005). Royalties for patents are a direct result of achievements from the development phase. Whether these patents are used in the later phases of the product life cycle is irrelevant as far as the generation of these revenues is concerned.

6.7.1.8 Market and service phase

Traditionally, the producers of technical goods relied solely on profits generated through the sale of their products. With today's decreasing sales profits and increasing service profits, such negligence becomes dangerous (Mateika 2005).

There are comprehensive models available for costs and revenues in the market and service phase seen from the perspective of the user. Mateika puts forward the proposition to analyse these models in the following way: all costs for the user can be seen as being potential sources of revenues for the manufacturer. He uses VDI 2884 as a basis for his selection of likely sources of revenues (Mateika 2005).

The same approach is adopted here with the difference that VDMA 34160 is used instead as the basis for the selection. This offers a "forecasting model for life cycle costs of machines and plants" seen from the perspective of the user (VDMA 34160 2006; Chapter 3.1.3).

	Potential Source of Revenues for:	
Cost Element	Other Party	Manufacturer
Warranty Extension	No	Yes
Training Costs	Yes	Yes
Construction/Conversion Costs	Yes	Yes
Costs for Network Infrastructure	Yes	Yes
Maintenance Flat Rate	No	Yes
Costs for unscheduled Repairs	No	Yes
Occupancy Costs	Yes	No
Energy Costs	Yes	No
Personnel Costs	Yes	Yes
...		

Table 6.8: Examples of potential sources of revenues during the market and service phase. (in accordance with VDMA34160)

Owing to the large number of cost elements named in VDMA 34160, only a few of them are given in the Table 6.8 above as an example of their suitability as a source of revenues for the manufacturer or other parties.

To assess the individual potential for a company, it is recommended that VDMA 34160 is consulted where all costs elements are listed and described. The provision of such services naturally creates costs at the same time. These costs have to be compared with the possible revenues in order to decide whether a certain service should be offered or not. Other services are regulated by law. Within the European Union, a warranty is given to private customers for two years from the date of delivery (EU 1999). In addition to that, the burden of proof remains with the seller of the product for at least the first six months from the delivery date. The consumer is entitled "to have the goods brought into conformity free of charge by repair or replacement, [...] or to have an appropriate reduction made in the price or the contract rescinded" (EU 1999). Although the manufacturer is not always the seller of the product, the chances of being made liable in some form by sellers or end consumers are high.

6.7.1.9 Disposal phase

As Mateika states, the disassembly, deconstruction and reuse of products can also create revenues for the manufacturer (Mateika 2005):

$$R_d = R_{da} + R_{dc} + R_{re} \qquad [5\text{-}16]$$

R_d	=	Revenues from disposal phase [€]
R_{da}	=	Revenues from disassembly [€]
R_{dc}	=	Revenues from deconstruction [€]
R_{re}	=	Revenues from reuse [€]

At the same time, there is a tendency to hold the manufacturer more and more responsible for the disposal costs for discarded products. For example, the EU has already passed legislation for the automobile and electronic industry which forces manufacturers to take back products at the end of their life cycle from customers and pay the costs of disposal (EU 2003 WEEE and EU 2000).

The VDMA 34160 includes a description of the costs after utilisation for the user (VDMA 2006). The extent to which the manufacturer is held liable for these costs depends on the current legislation. Here it is assumed that the manufacturer is not responsible for transportation, renovation or redevelopment but for the actual disposal of the product. The same costs emerge if the manufacturer buys or takes back the product voluntarily to extract reusable parts (except for costs of transport which may have to be taken into account in such a case). In both cases, the following formula [5-17] can be developed on the basis of VDMA 34160. Firstly, the product has to be dismantled and disassembled to decommission the machine and extract reusable parts. Parts which are not reused must be scrapped or disposed of:

$$C_{dp} = C_{dm} + C_{dd} + C_{sc} + C_{di} \qquad [5\text{-}17]$$

C_{dp}	=	Costs of the disposal phase [€]
C_{dm}	=	Dismantling costs [€]
C_{dd}	=	Disassembly and decommissioning costs [€]
C_{sc}	=	Scrapping costs [€]
C_{di}	=	Disposal costs [€]

This model describes the costs and revenues of the manufacturer throughout the life cycle of a product. Even more detail can be added to the model but, given the early stage of product development, it is difficult to further enhance the accuracy with reasonable effort. In the next paragraph, methods for the assessment and evaluation of these costs and revenues are discussed.

6.7.2 Costs and revenues analysis

The sold product can be seen as being an investment not only for the user but also for the manufacturer. For the manufacturer, the sale may be only the beginning of a long and profitable customer relationship. Services of all kinds can be offered to the customer, from the training of staff right up to the disposal or repurchase of the product at the end of its lifetime. As these revenues may be created in a distant future and may also become uncertain, it is necessary to take the time factor into consideration. There are several well-proven methods of capital budgeting which fulfil this requirement.

6.7.2.1 Methods of capital budgeting

Capital budgeting can be divided into static and dynamic methods. While dynamic methods take the moment in time of the emergence of costs into account, static methods neglect the chronology of costs and revenues. As products designed with attention to a life cycle perspective generally have a long lifetime and create high costs while in use, dynamic methods are used here to analyse the costs and revenues throughout the life cycle. Dynamic methods are superior to static methods in such cases because they take the effects of interest rates into account which become particularly important due to the characteristics described above (Schmidt

2000), (Mateika 2005). In the following, the net present value method, the internal rate of return method and the payback period rule are discussed as options to compare the profitability of different life cycle concepts.

6.7.2.2 Net present value method

This is one of the most common methods used to calculate the present value of net cash flows (NPV, see formula [5-18] below) by *discounting* future cash flows. The net cash flow of a period is the difference between the incoming and outgoing payments during that period. From the point of view of the manufacturer, examples of such payments in the later stages of the product life cycle could be cash inflows from additional services during the use of the product or costs for forced take-back at the end of the life cycle. In dependence upon Schäfer and Mateika, the following interpretation of the net present value of a sold product for the manufacturer is proposed (Schäfer 1999),(Mateika 2005):

$$NPV_m = C_{0i} - C_{0o} + \sum_{t=1}^{T-1}(C_{ti} - C_{to})\times(1+r)^{-t} + (C_{ie} - C_{oe})\times(1+r)^{-T}$$

[5-18]

NPV_m	=	Net present value for the manufacturer[€]
C_{0i}	=	Sales revenue [€]
C_{0o}	=	Cost of production [€]
C_i	=	Cash inflow during use in period t [€]
C_o	=	Cash outflow during use in period t[€]
r	=	Discount rate
t	=	Time of the cash flow in periods
T	=	Last period of the life cycle
C_{ie}	=	Cash inflow at the end of life cycle [€]
C_{oe}	=	Cash outflow at the end of life cycle [€]

If the actual interest rate for the life cycle concept is higher than the discount rate, the NPV_m is positive and the concept is implemented because it will create more revenues than an alternative investment. This makes the discount rate crucial for the result of the net present value method. The discount rate is chosen with regard to external and internal costs of capital and the risk of the investment in the life cycle concept (Hoffmeister 2000). The NPV_m method is especially advantageous for evaluating life cycle concepts which shift revenues for example from the sale of the product to later phases of the life cycle. The internal rate of return can be calculated on the basis of the NPV_m.

6.7.2.3 The internal rate of return method

With this method, the internal yield of an investment is calculated by setting the NPV_m equal to zero.

$$NPV_m = C_0 + \sum_{t=1}^{T-1}(C_i - C_o)\times(1+r)^{-t} + (C_{ie} - C_{oe})\times(1+r)^{-T} \overset{!}{=} 0$$

[5-19]

According to Schäfer, it can be assumed that an analytic solution is impossible if there are more than three time periods. In practice, an approximate solution can easily be calculated by using two trail interest rates and subsequently conducting a linear interpolation (Schäfer 1999). This approximate solution can then be compared with the interest rates which can be earned by applying alternative life cycle concepts. This makes it possible to select the life cycle concept with the highest rate of return. However, both the NPV_m method and the internal rate of return method neglect possible risks associated with long-term investments.

6.7.2.4 Payback period rule

Amortisation is reached when the sum of the return flow equals the original investment plus the interest rate. This method tries to evaluate the risk of an investment: the longer the period of time between investment and amortisation, the higher the risk of losing money through cash losses (Mateika 2005). Therefore, it is used as an additional evaluation method. It is only possible to compare two alternative life cycle concepts if the products involved have the same lifespan because the annual amortisation depends on the useful economic life (Schäfer 1999). According to Mateika, the amortisation time is located between the point in time T_s and T_{s+1} if (Mateika 2005):

$$\sum_{t=0}^{T_s}(R_t - C_t)\times r^{-1} < 0 \quad and \quad \sum_{t=0}^{T_{s+1}}(R_t - C_t)\times r^{-1} > 0 \qquad [5-20]$$

R_t	=	Revenues during the period t [€]
C_t	=	Costs during the period t [€]
t	=	Time of revenues and costs
r	=	Discount rate

T_s can be calculated by adding together the revenues and costs until the sum becomes positive. The point of time T_s is the moment before this happens and T_{s+1} is the one where the sum becomes positive for the first time (Schäfer 1999). An approximate solution can again be obtained using the linear interpolation of $NPVm_{Ts}$ and $NPVm_{Ts+1}$ (Mateika 2005).

Mateika stresses the significance of the methods of capital budgeting in the analysis stage of the life cycle-orientated product development (Mateika 2005). Long periods of amortisation underline the importance of the evaluation of costs and revenues in the later phases of the life cycle. Short periods diminish the value of life cycle-orientated product design in the eyes of the manufacturer. The internal rate of return method enables the manufacturer to gain an insight into the profitability of specific stages of the product life cycle.

These considerations complete the analysis of costs and revenues of life cycle concepts. However, there are many other factors which have an important influence on the economic success of a manufacturer. After all it is the customers who make purchase decisions and determine which competitor succeeds in the market.

6.7.3 Users

The importance of the integration of customers in the product development process is generally acknowledged. The question remains when and how this integration is supposed to take place.

Kohn and Niethammer have examined the premises under which the early integration of customers in the product development process is beneficial (Kohn and Niethammer 2005). They distinguish four different types of innovation. If the increase in performance or the reduction of costs is large, or new features are added, the innovation is *radical*. When this is not the case, the innovation can be called *incremental*. If the innovation is based on existing knowledge, it is called *sustaining*. When existing knowledge is rendered obsolete, the innovation is *disruptive*.

It is hard for current customers to think "out of the box", if they are asked to suggest improvements for current products. They mostly call for the linear enhancement of current capability characteristics (Kohn and Niethammer 2005). If the goal is to optimise existing products, customers can bring in valuable experience (Table 6.9). New life cycle concepts however will mostly require radical or disruptive innovations or include entering new markets. In this case the early integration of customers in the development process offers little advantages. Thus only complete life cycle concepts are presented to possible customers. This approach is supported by Kohn and Niethammer, who doubt that customer involvement in the creation and selection of ideas is advantageous (Kohn and Niethammer 2005). For these reasons, users are involved in the evaluation of life cycle concepts. Tools for assessing the preferences of customers have been developed on the field of marketing research. Marketing research "is the systematic process of gathering and analysing data for marketing decisions" (Hüttner and Schwartning 2002).

There are basically two different ways of gaining information: primary and secondary research. While secondary research includes the "provision, combination and interpretation of already existing knowledge" (Meffert 2000), primary research requires the "creation of data especially for the study" (Hüttner and Schwarting 2002). Information gained through secondary research from internal or external sources provides a basic understanding of underlying problems at relatively low cost and within a short period of time (Meffert 2000). If, for example, the implementation of a life cycle concept requires new markets to be entered, statistical data about the size of the market can be gained through secondary research. However, if more specific data are needed, primary research is indispensable.

Types of Innovation and Customer Integration		
Types of Innovation	Sustaining	Disruptive
Incremental	Early integration of customers beneficial	Early integration of customers often impossible, since customers are unknown
Radical	Early integration of customers often not beneficial, since customers are unable to think "out of the box"	Early integration of customers neither possible nor beneficial

Table 6.9: Types of innovation and customer integration. (Kohn and Niethammer 2005)

Hüttner and Schwarting name four different methods of data acquisition in primary research: survey, observation, experiment and panels (Hüttner and Schwarting 2002). The survey is the most widespread and important marketing method for gaining information (Meffert 2000). There are several different forms of surveys: written questionnaires (Meffert 2000), questionnaires which can be filled in online on the Internet (Berekhoven et al. 2004) and oral interviews conducted in the form of face-to-face interviews (Berekhoven et al. 2004) or by telephone (Meffert 2000).

Another important tool is the observation of customer behaviour. Meffert defines observation as "the systematic gathering of sensual perceivable facts at the time of their appearance" (Meffert 2000).

The probably least important instrument is experiments. An experiment can be seen as being "repeatable, under controlled conditions that are defined in advance conducted test, with which the measurement of the impact of one or several independent factors on the respective dependent factor is measured to test a hypothesis empiric" (Meffert 2000). From the categories of goods, the durable goods and services are important (Berekhoven et al. 2004).

Valuable insights into the chances of a new life cycle concept in the future may be gained through panels because the goal of panels is the "research on changes in markets respectively behaviour" (Meffert 2000). The need for observation is justified by Berekhoven et al. with the marked transformations in today's economy (Berekhoven et al. 2004). Berekhoven et al. define a panel as a "certain, constant, identical circle of respondents, which is surveyed at regular intervals for the same object" (Berekhoven et al. 2004).

Panels are categorised according to the respondents or the object of the investigation. As the objective of this section is to determine the preferences of the customers, the so-called *customer panels* (Hüttner and Schwartning 2002) are of interest here. A significant advantage of panels is that they provide ongoing market monitoring. Such a constant observation is necessary in order to be able to react fast enough to disturbing events, as explained in Chapter 6.5.2.

Apart from this advantage, Hüttner and Schwartning point out the methodological problems of panels (Hüttner and Schwartning 2002): the most important problem is the question whether the surveyed individuals are representative for the group which is supposed to be surveyed. For instance, within the context discussed here, the question may arise as to whether the panel should include possible future customers or current customers.

In addition to these methods, Berekoven et al. present several tests which are especially suitable for assessing life cycle concepts through users (Berekhoven et al. 2004). They describe a *concept test* during which the test person does not rate actual physical products but rather a product concept. This product concept could very well have the form of a life cycle concept, which can be described through a briefing as proposed by Berekoven et al. (Berekhoven et al. 2004). As a result, the evaluation of the product does not rely on the experience of the physical product but instead on the individual comprehension of the product idea.

The extent and reasons for the preference of one life cycle concept over another can be assessed using what Berekhoven et al. call a *test for preferences* (Berekhoven et al. 2004). This enables not only strengths and weaknesses to be determined but also allows for room for improvement. The life cycle concepts can now be altered according to the wishes of the customer. The last remaining criterion for the assessment of life cycle concepts is the environmental impact which can be assessed by conducting an LCA.

6.7.4 Ecological assessment through LCA

There are several goals which life cycle assessment (LCA) follows: it can be used to identify areas and measures for improvement or to assess and communicate the achieved advancements concerning the ecological friendliness of newly-developed products (Bullinger et al. 2003). The LCA can also be used to "study the environmental consequences of possible [future] changes between alternative product systems" (DIN EN ISO 140410). All of these goals are important in this context.

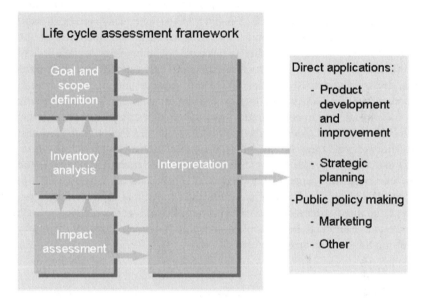

Fig. 6.16: Stages of an LCA. (DIN EN ISO 140410)

A new life cycle concept naturally still has a large potential for optimisation. At the same time, it is important to communicate the benefits of the new concept to customers and of course to achieve the objective set, i.e. the selection of a life cycle concept under the consideration of environmental friendliness. The ISO 14040 standard defines life cycle assessment as the "compilation and evaluation of the inputs, outputs and the potential environmental impacts of a product system throughout its life cycle"(DIN EN ISO 140410). The use of this standard offers several advantages. Results that rely on the same standard become comparable with each other and the evaluation process is also well-established and tested. "Any goods or services" (DIN EN ISO 140410) can be evaluated with the LCA, thus enabling the whole life cycle concept including all planned services to be assessed. For these reasons, the LCA according to ISO 14040 is recommended for the ecological assessment of life cycle concepts. In the following, the term LCA is applied within the context of the LCA as described by ISO 140410. The LCA has a life cycle perspective which includes all phases of the life cycle as well as interdependencies between them. It focuses only on the environmental impact of the product and not on other aspects such as economic or social implications. A functional unit, the "quantified performance of a product system, [is used] as a reference unit" (DIN EN ISO 140410) whose inputs and outputs are analysed. The LCA is an iterative technique, i.e. the results of one phase of the LCA are used in the others. Transparency and comprehensiveness are important features of the LCA in the same way as there is a preference for scientific methods based on natural science.

As shown in Figure 6.16, the results of the LCA and the interpretation of the results have implications for many fields. Here, the most important aspect is to further enhance the environmental friendliness of the product and to have a basis for the selection of a single life cycle concept. "Uncertainty and variability are often mentioned as factors complicating the interpretation of outcomes of LCAs" (Huijbregts 2002). While variability is the result of unpredictable variations in the real world, uncertainty is caused by simplifications and flaws during the conduction of the LCA (Huijbregts 2002). Given the diverse interactions between the environment and the increasing complexity of products, these difficulties are hard to overcome. An attempt to combine the assessment of costs and revenues throughout the life cycle with the LCA is made by Friedel and Osten-Sacken to achieve synergy effects (Friedel and Osten-Sacken 1998). However, such a process is hindered by the various indicators and dimensions (Friedel and Osten-Sacken 1998). Only time will tell if this interesting approach to reduce the effort of evaluation will become realisable. With the completion of the LCA, the evaluation of the life cycle concepts regarding the three criteria costs and revenues, users and environmental friendliness is concluded. The next paragraph is concerned with optimising the life cycle concepts on the basis of these evaluations.

6.7.5 Optimisation of life cycle concepts

Once the results of all the evaluation methods used are available, they can be compared with each other. Areas and requirements for improvements are uncovered and are used to further refine the life cycle concepts. After these refinements are made, the altered life cycle concept needs to be evaluated again to make sure that no change for the worse has taken place. From the interdependency matrix onwards, all areas that were subject to change have to be checked with regard to their influence on the earlier evaluation. This iterative procedure ensures that the findings of the evaluation are also used to improve the concepts, justifying the effort invested in the evaluation. To enhance the life cycle concepts, the various methods which concentrate on certain aspects of the life cycle of a product can be used, such as life cycle costing (see Chapter 3.1.1). Once the improvement of the life cycles has reached a satisfactory level, an executive summary can be prepared to present the life cycle concepts to the top management.

6.7.6 Selection of a life cycle design

The adoption of a new life cycle concept is a decision which is invariably made at a top management level. A new life cycle concept could include new business ideas and the entry in new business segments or even markets. Such decisions have a great strategic impact and are therefore made at the highest level of an organisation. The decision-making process is supported by the results of the life cycle design assessment. However, weighting of the three aspects of assessment is not included in the model for two reasons. First of all, the question how important the various aspects of the life cycle concept are can be answered very differently depending on the industry, image and many other factors influencing the position of a company. For example, a company with a "green" image which supplies an environmentally conscious market may find the results of the LCA crucial for its success. Another company supplying a business-to-business market may consider the ongoing costs of operation for their customers the essential motive to buy and thus rely heavily on the economic assessment. Not only the diversity of determining factors in different markets and industries argue against a weighting but also the nature of the decision-making process. It is unrealistic to think that a model calculating a single "perfect" solution will be readily adopted in commerce.

Discussions concerning a new life cycle concept are generally of a political character and have to reconcile the interests not only of the stakeholders inside but also outside the organisation (Espinosa-Orias and Sharrat 2006). If this decision-making process is to some degree rational, it will be made based on arguments delivered by the results of the assessment of the life cycle concepts.

Once the decision for a life cycle concept has been made, the requirements specifications can be developed. From this point onwards, conventional methods of product development can be used together with life cycle optimisation methods to guarantee the implementation of the life cycle concept as intended. Together with the implementation, it is essential to assess the extent to which the scenarios forming the basis of the life cycle concept actually occur. Discontinuities may require changes in the life cycle concept or even to stop the implementation altogether. The method for conducting ongoing monitoring is described in Chapter 6.5.2 and Chapter 6.7.3.

With this last step in selecting a single life cycle concept, the method for the design of life cycle concepts has reached its goal: the development of a well-crafted life cycle concept. The next chapter summarises the findings and procedures of the previous chapters and illustrates the possible results of the method.

6.8 Synthesis and exemplification

First of all, this chapter provides an overview of the method for the design of life cycle concepts. In the following, an existing business model is be used to exemplify the possible results of the method.

6.8.1 Synthesis of the method for the design of life cycle concepts

The first step of the method for the design of life cycle concepts is to conduct a scenario analysis. Influencing factors are determined based on the definition of the examination field as the phases of the life cycle. From these factors, the *relevant key factors* which have a significant influence either on the development or on other factors are selected. For each selected key factor, a descriptor, quantitative or qualitative parameter is chosen and its state in the future projected. To avoid contradictions between these descriptors, values of consistency are calculated. With the cluster analysis, groups with similar characteristics are formed as a basis for the formulation of the scenarios. The scenarios are put into the form of verbal descriptions representing the state of factors which influence the organisation, form markets and demands and are decisive for the success of the organisation. These descriptions can then be used for the method for the design of life cycle concepts. They are subject to ongoing monitoring in order to enable early and appropriate responses to possible changes to be made. With this, the first milestone is reached: the scenarios are recorded and can be used in the next step.

For the creation of new ideas, two major inputs are used. In addition to the scenarios, the fields of action are needed in order to provide a framework and to stimulate new ways of thinking about measures to close open life cycles or create new services. To break up existing patterns of thought, each field of action is reduced to a basic idea which represents an approach on an abstract level. This way it is possible to rethink existing ways and means of material recycling, remanufacturing and services and enables the milestone of defining the fields of action and allocating all life cycle-related measures to them to be achieved as illustrated in Figure 6.17.

The fields of action are confronted with the scenarios to spark new ideas. For each scenario, the question is asked: "How would material recycling/remanufacturing/services work if Scenario X occurs?" To include the perspective of the customer, a fourth field of action is added with the question: "What kind of product and services would customers need if Scenario X occurs?"

This permits a wide variety of ideas to be created which do not adhere to today's status quo and instead point in directions that have probably remained unexplored up till now by the specific industry.

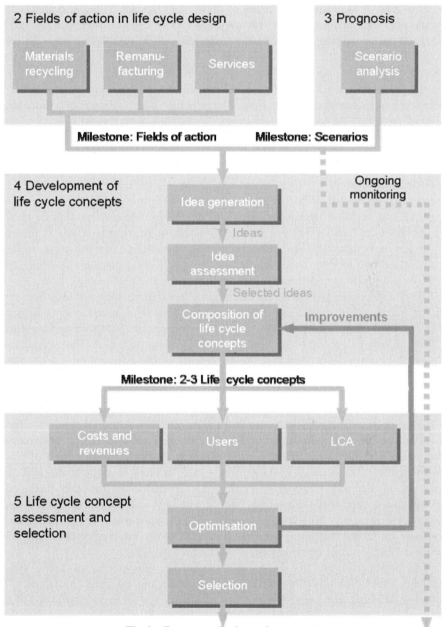

Fig. 6.17: Line of action of the method for the design of life cycle concepts.

However, at the same time, many of these ideas are unsuitable and unrefined. In the next step of idea assessment, they need to be selected and brought into a form which accurately represents the underlying idea.By shaping ideas in this way, it is possible to use the frequency of appearance as a criterion for the selection of ideas. An idea which appears more often than other ideas will probably be suitable for several possible futures and be thus more "future-proof". The second criterion for the selection of ideas is the value or benefit of an idea that is evaluated with the aid of the value benefit analysis. Using these two criteria, a selection can be made and the number of ideas reduced.

The remaining ideas form the basis for the composition of several life cycle concepts: fitting ideas are combined to create a life cycle concept. The interdependency matrix is then implemented to assess the coherence of a life cycle concept. If there are only few ideas, it is feasible to compute the sum of all positive and negative interdependencies of every possible combination of ideas. However, if the number of ideas is large, it makes more sense to compose the life cycle concepts first and then assess them in a second step. The outcomes of this process are two to three well-harmonised life cycle concepts, thus fulfilling the milestone of the step to develop two to three new life cycle concepts.

Three aspects are taken into consideration when making the final selection of a single life cycle concept: economic aspects (costs and revenues), ecological aspects (LCA) and customers (users). All three aspects are independent from each other and can be combined or omitted if they are out of place. The criteria are deliberately not weighted because the importance of the different criteria depends upon the situation of the organisation and industry concerned.

The costs and revenues of the three major phases of the life cycle of a product (development and manufacturing, market and service and disposal phase) are analysed in this early stage in as much detail as possible. Although these numbers are estimates, they remain important for the control of the actual costs which arise during implementation of the life cycle concept. Given the long period of time between the emergence of costs - for instance through manufacturing and the creation of revenues through services - it is necessary to take time into consideration. For this reason, the sale of the physical product is seen as an investment not only for the user but also for the manufacturer. At this stage, methods of capital budgeting can be utilised to assess the profit generated by a product throughout its entire life cycle for the manufacturer.

The perspective of the user is also crucial for the success of a life cycle concept: after all, it is the customers who make the purchase decisions. To appraise market acceptance for a new life cycle concept, the well-established tools of marketing research are used. While basic information can be gained through secondary research, more detailed and specific questions require primary research. This includes the acquisition of data, for example through surveys or experiments. Panels are utilised for the ongoing monitoring of certain developments which are decisive for the success of a life cycle concept.

Another important motivation for the creation of a new life cycle concept is environmental friendliness. LCA is used to evaluate the actual environmental burden created by a life cycle concept.

Once all the results from the evaluations are available, weaknesses in and possible improvements to the life cycle concepts become visible. If changes are made, it is essential that the changed life cycle concept is evaluated from the interdependency matrix onwards with regard to these alterations. This ensures that the objectives of the alterations made are actually met without having a serious negative influence on other life cycle measures. At this point, executive summaries of the life cycles can be prepared and presented to the top management. Based on the data at hand, the top management is then in a position to select a single life cycle concept, thus attaining the final milestone. By way of examples, the following section explains the results of the different stages of the method for the design of life cycle concepts.

6.8.2 Exemplification of the method

No practical examples of the use of the method for the design of life cycle concepts are available as yet. For this reason, a historic situation is used to further illustrate the implementation and possible outcomes of the method. Although it is difficult to reconstruct in retrospect the information available at a certain point of time and to blank out the actual situation of today, a self-conceived example set in the future is probably even less realistic. The goal of this paragraph is not to prove the applicability of the method but rather to illustrate it. Due to the scope of the method for the design of life cycle concepts, most steps are outlined without further explanation. The focus is on the more abstract outcomes of certain steps which might not have become clear in the synthesis. Try to imagine the situation of a producer of large copiers for business clients at the beginning or middle of the eighties in the last century. To improve and ensure the company's economic situation for the future and also to make products and processes more environmental friendly, a scenario analysis of possible future situations is conducted. Three of the resulting scenarios are given here an example:

Scenario 1 describes the "paper-free" office of the future where all documents are written, used and transferred electronically. Although this future has no space for a producer of copiers, it has to be considered in order to detect such a development early enough to enable an appropriate reaction. Thus, this scenario has to be subjected to ongoing monitoring.

Scenario 2 pictures the replacement of typewriters by computers with a certain amount of documentation on paper and written, non-electronic communication remaining. This creates a demand for output equipment and plotting devices for business clients together with a continued demand for copiers.

Scenario 3 depicts the triumph of the personal computer that also brings about a demand for output equipment. However, as this equipment is not only required by large business clients but also by private customers, it is therefore decentralised: each PC has its own printer.

Matrix for the Collection of Ideas				
Scenarios Fields of Action	Scenario 1	Scenario 2	Scenario 3	...
Material Recycling	- None	- Housing	- Housing - Plastic material	
Remanufacturing	- None	- Platen		
Services	- None	- All- inclusive Contract: customer pays per copied/printed page	- Extended Warranty - Leasing Contract	
Customers	- None	- Multi- functional	- Multi- functional	

Table 6.10: Example of a matrix for the collection of ideas.

By confronting these and other scenarios with the three fields of action (material recycling, remanufacturing and services plus the field of customers), the following ideas and many more are created by using creativity techniques (Table 6.10).

Once the ideas have been roughly sorted according to their frequency of appearance, they are assessed according to their value with the aid of the value benefit analysis. Using the interdependency matrix, the following two life cycle concepts are found to be consistent:

The cornerstone of the life cycle concept "full service for business clients" is formed by large, durable copiers which are maintained by the manufacturer. Several stages of service are offered up to the last stage, where the customer only pays a fixed price for every copy made. Through the indirect sale of paper and toner, the manufacturer enters new markets. As the manufacturer merely sells a function to the customer and not a physical product, it is much easier to implement remanufacturing measures. Resistance to products with remanufactured parts can be reduced significantly if the responsibility of the manufacturer is sustained throughout the product's life cycle. Remanufacturing and maintenance measures are simplified by a modular concept. Such a complex offer is only profitable for larger clients who require a high number of pages per day.

The other life cycle concept is the "small printer next to the PC". The basic idea behind this concept is to enter a new market and for this, the existing printing know-how of the company can be used. This concept addresses both large business clients and private customers. A low price is required to combat the reluctance of private customers to pay as well as to compensate for the large number of printers needed by business clients. The wide distribution of the printers and the necessary abandonment of valuable parts make material recycling the only consistent possibility to close open life cycle loops. Due to the too-high costs, elaborate services such as those included in the "full service for business client" concept cannot be offered. Basic services are thus provided by retailers and the manufacturer enters no new markets through the provision of services.

The assessment of the costs and revenues of the life cycle concepts show that it is economically impossible to build large series of small printers with the equipment and staff currently available and that it would require massive investment to make this production profitable. The concept "full service for business clients" can be easily applied with minor changes due to the modular concept. The implementation of services requires only little personnel and resources right at the beginning.

During the assessment of the preferences of both business clients and private users, it becomes obvious that there is a demand for multi-functional products which combine the functions of a copier with a printer, fax or other applications. However, the effort required to develop such a product in a compact form and the reluctance of private customers to pay for them make this concept a solution which is only feasible for business clients. As a result, this idea is used in the concept "full service for business clients".

The LCA shows that the concept "small printers next to the PC" produces a much larger amount of waste than the life cycle concept "full service for business clients". On the basis of this evaluation, it could well be imagined that the life cycle concept "full service for business clients" is chosen.

However, let us return to actual developments in the real world. In 1991, Hildebrandt described the business model of a small company which remanufactures copiers for manufacturers and retailers. When the company started to remanufacture copier modules, "environmental protection, ecology and product recovery were foreign words within the industry sector" (Hildebrandt 1991). To gain acceptance, an economically interesting concept was essential (Hildebrandt 1991). When it became obvious that the model was profitable, other companies also started to remanufacture copiers. As Hildebrandt states, the initial investment in a copier can be reduced through remanufacturing. This leads to shorter recovery periods which are beneficial for the provider if the user pays a fixed price for each copy made (Hildebrand 1991). This is a good example of positive interdependencies in a life cycle concept (see Chapter 6.3.3). In 1999, Üffinger reported about the remanufacturing of copiers (Üffinger 1999). By then, the concept seemed to have spread to the large manufacturers in the industry: Üffinger was speaking as a representative of the Kodak AG (Üffinger 1999).

Today, the main ideas of the life cycle concept "full service for business clients" described above can be found for example at the company Nashuatec in Germany. The following description of Nashuatec's business model is based on the information available on their homepage (Nashuatec 2008). Multi-functional products which combine several of the following functionalities are offered as well as copiers, printers, scanners and fax machines. Customers can chose between several kinds of maintenance contract. In addition to these basic services, Nashuatec provides consultation regarding office communication and the management of documents, training courses and project management services. If a customer chooses the service "1=1 PAY PER PAGE", they simply rent the machines and pay a single fixed price for each page that is printed, copied or faxed. Nashuatec was the first company to provide such a service in 1996 and has become an example for many similar offers (Komplettlösungen rund ums Drucken plus Dokumenten-Management 2002).

Nashuatec collects empty toner cartridges at no charge for the user to reuse them. Awards for environmental friendliness are named on the website and the efforts for the environment are described to create a positive image of the company. Nashuatec claims to design its products to enable and simplify remanufacturing and recycling, for example by standardising plastic materials.

6.9 Conclusion and outlook of the methodological approach

Based on the situation in which manufacturers find themselves today, the current gap in research is formulated in Chapter 6.2 and 6.3. Both the ecological and economical motivations behind taking life cycles into consideration are explained and their possible synthesis is proposed. This concretises requirements for the development of life cycle concepts.

In Chapter 6.4, the measures to close open loops in material flows and to *dematerialise* the products manufactured by a company are classified into three fields of action. While material recycling and remanufacturing aim at closing open loops in the flow of materials, it is possible to dematerialise through the provision of services.

Chapter 6.5 focuses on generating descriptions of possible future situations. From several possible methods, the scenario analysis is chosen to create these descriptions. The reason for this selection is the unsuitability of the other methods and the applicability of the drafted scenarios for the idea generation process. The scenarios also have another important function: the actual state of things is constantly analysed in order to detect the occurrence of certain scenarios early enough to enable appropriate measures to be taken.

These "pictures of the future" in the form of scenarios are then also used in Chapter 6.6 to find new ideas for life cycle concepts in the fields of action outlined in the classification. Various creativity techniques are proposed to facilitate this search for ideas.

Once enough ideas have been created, the most promising and valuable ideas are determined. In a first step, the frequency of appearance and value benefit analysis are used to pre-select the ideas. These are then used to develop several coherent life cycle concepts. With the aid of an interdependency matrix, the interdependencies between different ideas are taken into account and coherence is assured. These new life cycle concepts include not only a product definition but also a description of the whole life cycle of the product from its manufacture right up to its disposal or remanufacturing as well as all associated flows of material.

To determine the most appropriate life cycle concept, all are assessed with regard to three criteria: profitability, market and environmental friendliness. These are described in detail in Chapter 6.7. The criteria are not weighted because their importance may vary considerably from one company to another. This makes it possible to use only the criteria which are relevant to the decision-making process in each individual case.

The first criterion is formed by the costs and revenues created by a specific life cycle concept. The complete life cycle is analysed in this respect. To evaluate these costs and revenues, the factor of time is taken into account using capital budgeting methods. The next group of criteria are user preferences and these are assessed by applying marketing research methods. Panels are especially suitable for monitoring changes in user preferences. In a final step, the environmental friendliness of the concepts is assessed by conducting an LCA.

Naturally, some life cycle concepts will show need or room for improvement at the end of the evaluation process. After making such alterations, it is necessary to evaluate the life cycle concept again with respect to the outcomes of these changes. On the basis of these evaluations, the top management can then decide on a single life cycle concept. In Chapter 6.8, the method itself is outlined in the synthesis and is explained using a historic example.

6.10 References concerning chapter 6

(Backhaus 1997) Backhaus, K. 1997. *Industriegütermarketing.* München, Germany: Verlag Franz Vahlen GmbH.
(Beck et al. 2000) Beck, K., Glotz, P. and Vogelsang, G. 2000. Die Zukunft des Internet: Internationale Delphi-*Befragung zur Entwicklung der Online-Kommunikation.* Konstanz, Germany: UVK Medien Verlagsgesellschaft
(Berekhoven et al. 2004) Berekhoven, C., Eckert W. and Ellenrieder, P. 2004. *Marktforschung: Methodische Grundlagen und praktische Anwendung.* Wiesbaden, Germany: Gabler Verlag.
(Bullinger et al. 2003) Bullinger, H.-J., Warnecke, H.J. and Westkämper, E. (ed.) 2003. *Neue Organisationsformen im Unternehmen.* Berlin, Germany: Springer Verlag.
(Czichos et al. 2004) Czichos, H., Hennecke, M. and Akademischer Verein Hütte e.V. (ed.) 2004. *Hütte: das Ingenieurswissen.* Berlin, Germany: Springer Verlag.
(Czichos et al. 2008) Czichos, H. and Hennecke, M. 2008. *Hütte: Das Ingenieurswissen.* Berlin: Springer Verlag.
(DIN 1995) DIN- Deutsches Institut für Normung e.V. 1995. *DIN EN ISO 8402 Quality management and quality assurance: Vocabulary.* Berlin, Germany: Beuth Verlag.
(Espinosa-Orias and Sharrat 2006) Espinosa-Orias, N. and Sharrat, P. N. 2006. A hierarchical approach to stakeholder engagement. *Proceedings of the 13th International Seminar on Life Cycle Engineering CIRP,* Leuven, 31 May- 1 June 2006.
(EU 1999) EU. 1999. Directive 1999/44/EG of the European Parliament and of the Council of 25 May 1999 on certain aspects of the sale of consumer goods and associated guarantees. *Official Journal of the European Union,* L 171: 12-16, July 7.
(EU 2000) EU. 2000. Directive 2000/53/EC of the European Parliament and of the Council of 18 September 2000 on end-of life vehicles. *Official Journal of the European Union,* L 269: 34-42, October 21.
(EU 2003)EU. 2003. Directive 2002/96/EC of the European Parliament and of the Council of 27 January 2003 on waste electrical and electronic equipment (WEEE). *Official Journal of the European Union,* L 037: 24-39, February 13.
(Eversheim and Schuh 1999) Eversheim, W. and Schuh, G. (ed.) 1999. *Integriertes Management.* Berlin, Germany: Springer Verlag.
(Federal Statistical Office of Germany 2007) Federal Statistical Office of Germany. 2007. *Services, financial services.* [Online] Available: http://www.destatis.de/jetspeed/portal/cms/Sites/destatis/Internet/EN/Navigation/Statitics/DienstleistungenFinanzdienstleistun gen/DienstleistungenFinanzdienstleistungen.psml [accessed 17 November 2007]
(Federal Statistical Office of Germany 2008a) Federal Statistical Office of Germany. 2008a. *Preise und Preisindizes für gewerbliche Produkte (Erzeugerpreise) - Fachserie 17 Reihe 2 - Januar 2004.* [Online] Available: https://www-ec.destatis.de/csp/shop/sfg/bpm.html.cms.cBroker.cls?cmspath=struktur,vollanzeige.cspandID=1012671 [accessed 7 January 2008]
(Federal Statistical Office of Germany 2008b) Federal Statistical Office of Germany. 2008b. *Preise und Preisindizes für gewerbliche Produkte (Erzeugerpreise) - Fachserie 17 Reihe 2 - Januar 2005.* [Online] Available: https://www-ec.destatis.de/csp/shop/sfg/bpm.html.cms.cBroker.cls?cmspath=struktur,vollanzeige.cspandID=1015871 [accessed 7 January 2008]
(Federal Statistical Office of Germany 2008c) Federal Statistical Office of Germany. 2008c. *Preise und Preisindizes für gewerbliche Produkte (Erzeugerpreise) - Fachserie 17 Reihe 2 - Januar 2006.* [Online] Available: https://www-ec.destatis.de/csp/shop/sfg/bpm.html.cms.cBroker.cls?cmspath=struktur,vollanzeige.cspandID=1018009 [accessed 7 January 2008]

(Federal Statistical Office of Germany 2008d) Federal Statistical Office of Germany. 2008d. *Preise und Preisindizes für gewerbliche Produkte (Erzeugerpreise) - Fachserie 17 Reihe 2 - Januar 2007.* [Online] Available: https://www-ec.destatis.de/csp/shop/sfg/ bpm.html.cms.cBroker.cls?cmspath=struktur,vollanzeige.csp&ID=1019951 [accessed 7 January 2008]

(Friedel and Osten-Sacken 1998) Friedel, A. and Osten-Sacken, v.d.D. 1998. Kombinierte Lebenslauf-Erfolgsrechnung und Ökobilanzierung. *Konstruktion: Zukunft für Produktentwicklung und Ingenieurwerkstoffe.* 6: 40-44.

(Gausemeier et al. 1995) Gausemeier, J., Fink, A. and Schlake, O. 1995. *Szenario-Management: Planen und Führen mit Szenarien.* München, Germany: Hanser Verlag.

(Gausemeier et al. 1996) Gausemeier, J., Fink, A. and Schlake, O. 1996. Scenario-management during the early stages of product development. *Proceedings of the IFIP WG5.3 international conference on Life-cycle modelling for innovative products and processes,* Berlin, Germany, November/December 1995. 369-380. London: Chapman and Hall.

(Gordon 1961) Gordon, W.J.J. 1961. Synectics: the development of creative capacity. New York: Harper and Row.

(Hahner 2000) Hahner, C. A. 2000. Bewertung von Innovationsideen mit Hilfe von Lebenszyklusaufwand. Stuttgart, Germany: University of Stuttgart. (Dissertation).

(Hewlett Packard 2007a) Hewlett Packard. 2007a.HP Druckerpatrone Schwarz (C9351 AE): Spezifikationen. [Online] Available at: http://h10010.www1.hp.com/wwpc/de/de/ ho/WF06c/A1-14607-14709-14709-14787-12203906-45287899.html [accessed 24 November 2007]

(Hewlett Packard 2007b) Hewlett Packard. 2007b.HP Deskjet D 2460 Farbtintenstrahldrucker (CB611A): *Spezifikationen.* [Online] Available at: http://h10010.www1.hp.com/wwpc/de/de/ho/WF06b/2923-2929-3289-3289-12429406-80080343-80109798.html?jumpid=reg_R1002_DEDE [accessed 24 November 2007]

(Hildebrandt 1991) Hildebrandt, K. 1991. Produktrecycling von Kopiergeräten. *VDI Bericht 906: Recycling eine Herausforderung für den Konstrukteur,* Bad Soden, 14-15 November 1991. 143-151. Düsseldorf, Germany: VDI-Verlag.

(Hill 2005) Hill, B. 2005. 'Naturorientierte Innovationsstrategie: Entwicklen und Konstruieren nach biologischen Vorbildern'; in *Bionik: Aktuelle Forschungsergebnisse in Natur-, Ingenieur- und Geisteswissenschaft.* Rossmann, T. and Tropea, C. (ed.). Berlin, Germany: Springer.

(Hoffmeister 2000) Hoffmeister, W. 2000. *Investitionsrechnung und Nutzwertanalyse.* Stuttgart: Verlag W. Kohlhammer.

(Holzbaur 2007) Holzbaur, U. 2007. Entwicklungsmanagement: Mit hervorragenden Produkten zum *Markterfolg.* Berlin, Germany: Springer Verlag.

(Huijbregts 2002) Huijbregts, M. 2002. Uncertainty and variability in environmental life-cycle assessment. Amsterdam, Netherlands: University of Amsterdam (Dissertation).

(Hüttner and Schwarting 2002) Hüttner, M. and Schwarting, U. 2002. *Grundzüge der Marktforschung.* München, Germany: R. Oldenbourg Verlag.

(Kimura and Suzuki 1996) Kimura, F. and Suzuki, H. 1996. Design of Right Quality Products for Total Life Cycle Support. *Proceedings of the 3rd International Seminar on Life Cycle Engineering CIRP,* Zürich, 18-20 March 1996. 127-133. Tokyo, Japan: Department of Precision Machinery Engineering.

(Kobayashi 2003) Kobayashi, H. 2003. Idea generation and risk evaluation methods for life cycle planning. Proceedings of EcoDesign2003: Third International Symposium on Environmentally Conscious Design and Inverse Manufacturing, Tokyo, 8-11 December. 117-123.

(Kohn and Niethammer) Kohn, S. and Niethammer, R. 2005. Kundeneinbindung in den Innovationsprozess. *In:* Barske, H., ed. *Innovationsmanagement.* CD-ROM. Düsseldorf, Germany: Symposium Publishing.

(Kölscheid 1999) Kölscheid, W. 1999. Methodik zur lebenszyklusorientierten Produktgestaltung: Ein Beitrag *zum Life Cycle Design*. Aachen, Germany: RWTH Aachen. (Dissertation). Shaker Verlag.

('Komplettlösungen rund ums Drucken plus Dokumenten-Management' 2002) 'Komplettlösungen rund ums Drucken plus Dokumenten-Management' 2002. *BIT Business Information Technology*, No. 6, February 2002. [Online] Available at: http://www.bitverlag.de/bitverlag/artikel/index.asp?ogr=bit&such=nrg&log=and&fra=1 &frp=0&item=1107#top [accessed 19 Januar 2008]

(Lindahl et al. 2005) Lindahl, M., Sundin, E., Östlin, J. & Björkman, M. 2005. Concepts and definitions for product recovery – Analysis and clearification of the termininology used in academia and industry. *Proceedings of the 2005 CIRP Seminar on Life Cycle Engineering*, Grenoble, France, 3-5 April 2005. 123-138.

(Mateika 2005) Mateika, M. 2005. Unterstützung der lebenszyklusorientierten Produktplanung am Beispiel *des Maschinen- und Anlagenbaus*. Braunschweig, Germany: University of Braunschweig (Dissertation).

(Meffert 2000) Meffert, H. 2000. Marketing: Grundlagen marktorientierter Unternehmensführung. Wiesbaden, Germany: Gabler Verlag.

(Meier 2003) Meier, H. 2003. Dienstleistungsorientierte Geschäftsmodelle im Maschinen- und Anlagenbau. Berlin, Germany: Springer Verlag.

(Mercedes-Benz Bank AG 2008) Mercedes-Benz Bank AG. 2008. *Historie der Mercedes-Benz Bank AG*. [Online] Available at: http://www.mercedes-benz-bank.de/intrade/cms/ UN_UESHI_Uebersichtsseite_Historie.html?linkArea=mainnavi2 [accessed 18 January 2008]

(Nashuatec 2008) Nashuatec. 2008. *Nashuatec Deutschland*. [Online] Available at: http://www.nashuatec.de/ [accessed 16 January 2008]

(Niemann 2006) Niemann, J. 2006. Die Kosten stets im Griff. *VDMA Nachrichten*, no. 5, p. 41-42.

(Niemann 2007) Niemann, J. 2007. Eine Methodik zum dynamischen Life Cycle Controlling von *Produktionssystemen*. Stuttgart, Germany: University of Stuttgart (Dissertation). Heimsheim, Germany: Jost-Jetter.

(Ophey 2005) Ophey, L. 2005. Entwicklungsmanagement: Methoden in der Produktentwicklung. Berlin, Germany: Springer Verlag.

(Osten-Sacken 1999) Osten-Sacken, v. d. D. 1999. Lebenslauforientierte, ganzheitliche Erfolgsrechnung für *Werkzeugmaschinen*. Stuttgart, Germany: University of Stuttgart (Dissertation).

(Pahl et al. 2007) Pahl, G., Beitz W., Feldhusen J. and Grote, K.-H. 2007. *Konstruktionslehre: Grundlagen erfolgreicher Produktentwicklung: Methoden und Anwendung*. Berlin, Germany: Springer Verlag.

(Schäfer 1999) Schäfer, H. 1999. Unternehmensinvestitionen: Grundzüge in Theorie und Management. Heidelberg, Germany: Physica-Verlag.

(Schmidt 2000) Schmidt, F. R. 2000. *Life Cycle Target Costing*. Leipzig, Germany: University of Leipzig. (Dissertation).

(Schnetzler 2004) Schnetzler, N. 2004. *Die Ideenmaschine*. Weinheim: Wiley-VCH Verlag

(Senti 1994) Senti, R. 1994. Produktlebenszyklusorientiertes Kosten- und Erlösmanagement. St. Gallen, Switzerland: University of St. Gallen (Dissertation).

(Spath and Schuster 2004) Spath, D. and Schuster, E. 2004. 'Neue Geschäftsmodelle durch Betreibermodelle', *wt Werkstattstechnik-online*, vol. 7/8-2004, p. 319-321 [Online] Available at: http://www.werkstattstechnik.de/wt/index.php [accessed 8 January 2008]

(Specht and Beckmann) Specht, G. and Beckmann, C. 1996. *F&E-Management*. Stuttgart, Germany: Schäfer-Poeschel.

(Steinhilper 1998) Steinhilper, R. 1998. Remanufacturing: The Ultimate Form of Recycling. Stuttgart, Germany: Fraunhofer IRB Verlag.

(Stevels 2000) Stevels, A. 2000. Five ways to make money while being green. *Proceedings of the 7th International Seminar on Life Cycle Engineering CIRP*, Tokyo, Japan, 27-29 November 2000. 23-29

(Tomiyama et al. 1996) Tomiyama, T., Sakao T. and Umeda Y. 1996. The post-mass production paradigm, knowledge intensive engineering, and soft machines. *Proceedings of the IFIP WG5.3 international conference on Life-cycle modelling for innovative products and processes*, Berlin, Germany, November/December 1995. 369-380. London: Chapman & Hall.

(Tomiyama et al. 1997) Tomiyama, T., Umeda, Y. and Wallace, D. R. 1997. A holistic approach to life cycle design, in *Life Cycle Networks*, ed. Krause, F.-L. and Seliger (eds.) London: Chapman & Hall.

(Üffinger 1999) Üffinger, G. 1999. Remanufacturing: am Beispiel von Kopiergeräten. *Fraunhofer IPA-Technologie-Forum on Life Cycle Management*, Stuttgart, 29 April 1999. 49-54. Stuttgart, Germany: FpF – Verein zur Förderung produktionstechnischer Forschung.

(Umeda et al. 2000) Umeda, Y., Nonomura, A. and Tomiyama, T. 2000. Study on life-cycle design for the post mass production paradigm. *Artificial Intelligence for Engineering Design, Analysis and Manufacturing* 14(2): 149-161, April.

(VDI-Verein deutscher Ingenieure 1990) VDI-Verein deutscher Ingenieure. 1990, VDI 2860: Assembly and handling: Handling *functions, handling units Terminology, definitions and symbols*. Düsseldorf, Germany: Verein deutscher Ingenieure.

(VDI 1993) VDI-Verein deutscher Ingenieure 1993, VDI 2221: Methodik zum Entwickeln und *Konstruieren technischer Systeme und Produkte*. Düsseldorf, Germany: Verein deutscher Ingenieure.

(VDI 2003) VDI-Verein deutscher Ingenieure 2003, VDI 2884: Beschaffung, Betrieb und Instandhaltung *von Produktionsmitteln unter Anwendung von Life Cycle Costing (Entwurf)* Düsseldorf, Germany: Verein deutscher Ingenieure.

(VDMA 2002) VDMA-Verein Deutscher Maschinen- und Anlagenbauer e.V. 2002, Prozesse beschleunigen und gewinnorientiert steuern: Empfehlungen zur Unternehmensführung in der Investitionsgüterindustrie. Frankfurt/Main, Germany: VDMA Verlag

(VDMA 2006) VDMA-Verein Deutscher Maschinen- und Anlagenbauer e.V. 2006, *VDMA 34160: Forecasting Model for Lifecycle Costs of Machines and Plants*. Berlin, Germany: Beuth Verlag.

(Westkämper and Niemann 2006) Westkämper, E. and Niemann J. 2006. Dynamic Life Cycle Performance Simulation of Production Systems. *Proceedings of the 2005 CIRP Seminar on Life Cycle Engineering*, Grenoble, France, 3-5 April 2005. 419-428.

(Westkämper 2003) Westkämper, E. 2003. Assembly and Disassembly Processes in Product Life Cycle Perspectives. *CIRP Annals Manufacturing Technology*, no. 2, p. 579-588.

(Westkämper 2006) Westkämper, E. 2006. Wertschöpfung im Lebenszyklus der Produkte. *VDMA Nachrichten*, no. 5, p. 24.

(Zehbold 1996) Zehbold, C. 1996. *Lebenszykluskostenrechnung*. Wiesbaden, Germany: Gabler Verlag.

(Zeller 2003) Zeller, A. 2003. *Technologiefrühaufklärung mit Data Mining*. Stuttgart, Germany: University of Stuttgart. (Dissertation).

(Züst 1996) Züst, R. 1996. Sustainable Products and Processes. *Proceedings of the 3rd International Seminar on Life Cycle Engineering CIRP*, Zürich, 18-20 March 1996. 5-10. Zürich, Switzerland: Institute for Industrial Engineering and Management.

Internet Investigation: Prices for printers:

Amazon. 2007. *Price for EPSON Stylus DX4400*. [Online] Available at: http://www.amazon.de/
    Multifunktionsger%C3%A4t-Farbtintenstrahl-Drucker-Kopierer-Scanner/dp/tech-data/
    B000RARU06/ref=de_a_smtd?ie=UTF8andqid=1196075857&sr=1-31 [accessed 24
    November 2007]
Amazon. 2007. *Price for Hewlett Packard Deskjet D2460*. [Online] Available at:
    http://www.amazon.de/Hewlett-Packard-Deskjet-D2460-
    Tintenstrahldrucker/dp/B000RF9JGE/ref=sr_1_9/028-3072647-
    8823737?ie=UTF8&s=ce-de&qid=1196075766&sr=1-9 [accessed 24 November 2007]
Amazon. 2007. *Price for Canon Pixma MP510*. [Online] Available at:
    http://www.amazon.de/Canon-1450B006-Pixma-MP510-
    Multifunktionsger%C3%A4t/dp/tech-
    data/B000I5037M/ref=de_a_smtd?ie=UTF8&qid=1196077143&sr=1-7 [accessed 24
    November 2007]

Prices for cartridges:

Epson. 2007. *Price for cartridge for EPSON Stylus DX4400*. [Online] Available at:
    http://www.epson-
    store.de/cd/ld/content/shop/consumables/offer.php/sku/C13T07114010/devicesku/C11C
    688302 [accessed 24 November 2007]
Hewlett Packard. 2007. *Price for cartridge for Hewlett Packard Deskjet D2460*. [Online]
    Available at: http://h10010.www1.hp.com/wwpc/de/de/ho/WF29a/2923-2929-3289-
    3289-12429406-80080343.html?jumpid=reg_R1002_DEDE [accessed 24 November
    2007]
Amazon. 2007. *Price for cartridge for Canon Pixma MP510*. [Online] Available at:
    http://www.amazon.de/Canon-BJ-PGI-5BK-Tintenpatrone-
    schwarz/dp/B000B8TIZE/ref=pd_bbs_sr_1?ie=UTF8&s=ce-
    de&qid=1196077278&sr=8-1 [accessed 24 November 2007]

7 Summary

The development of modern products is being decisively influenced by the application of technologies which contribute towards increased efficiency. Products are becoming complex highly-integrated systems with internal technical intelligence, enabling the user to implement them reliably, economically and successfully even in the fringe ranges of technology. As a result, business strategies are aiming more and more towards perfecting technical systems, optimising product utilisation and maximising added value over the entire lifetime of a product. In this context, the total management of product life cycles associated with the integration of information and communication systems is becoming a key success factor for industrial companies.

Today, global competition demands guarantees from technical equipment manufacturers as far as the technical and economical performance of their products is concerned. This implies that machines have to work properly over long periods of time and perform different work tasks. To meet contract liabilities, machine manufacturers monitor their facilities, gather all manufacturing information and try to forecast and boost machine performance using intelligent process optimisation. Life cycle-orientated platforms provide data sources and experiences from other machines to supply information in order to reach excellence in manufacturing. Due to the increasing substitution of mechanical components for software, it has become possible to read out data from machines and process control in-situ, i.e. during a process. In this way, the current status of the machine can be called up, monitored and analysed online from anywhere in the world.

Modern technologies make it possible to couple actual process parameters with process data from the past, thus enabling the later process behaviour to be forecast in advance in a simulation. The early recognition of inefficiencies and faults allows in-situ process optimisation even in the absolute fringe ranges of precision and performance. These enormous potentials can be translated into costs and virtual profit information so that the processes can be "controlled" with regard to their economical efficiency. The transparent machine is becoming reality!

In modern system concepts, installations are operated in a distant network by tele-operations. It is possible that the installation may remain the property of the manufacturer and stay within his sphere of responsibility for its entire service life and that only the benefits or functions of the machine are sold. By integrating the knowledge sources of equipment manufacturers, machine outfitters and others, a markedly higher exploitation of the machine can be achieved.

To describe the maximum utilisation of products as product profit is the greatest paradigm change as it breaks away from the traditional paradigms of growth- and resource-optimisation. The emphasis is now being placed on customer usage or long-term business relationships and the focus is therefore on life cycles. This results in a new dimension, as by considering the concept of customer lifetime value, the worth of a business relationship can be determined with the aid of a dynamic life cycle model. In the case of the customer lifetime value concept, the capital value of payments associated with such a relationship is seen as a business relationship assessment criterion. This new viewpoint is pursued by worldwide networks which have come into existence to provide continuous product optimisation support. As a result, the performance capability of a company will be expressed in the future by its short-term access to such networks, which must be used to constantly optimise product usage in order for these companies to survive. Life cycle data are integrated into life cycle modelling and simulation, enabling users to analyse product life cycles systematically and evaluate them based on life cycle knowledge in an early phase of design activity and throughout the entire usage phase of the product. This overall application of such technologies provides an approach to master and permanently (re-design) optimised product life cycles.

The numerous concepts and examples from research and practice which are presented in the book demonstrate that "thinking in product life cycles" is no longer a vision but rather an economical, ecological and societal challenge with huge potentials. However, the successful design of sustainable product life cycles can only be achieved through the integrated consideration and participation of all life cycle partners to constantly optimise the product "from the cradle to the grave"

Index

Printing: Krips bv, Meppel, The Netherlands
Binding: Stürtz, Würzburg, Germany